Animals & Men

A positive breakthrough in Blue Dog research? Tasmania 2016 expedition report; Mark A Hall Obituary; A Mystery Carcass in Djibouti; The Gorilla – Does it exist?; in Search of Bowheads and Narwhals; Weird Weekend 2016; News, Reviews and More

The Journal of the Centre for Fortean Zoology

Animals & Men

Phantom Black Dogs in The Netherlands

PLUS: Tasmania 2017 Expedition Report; Phantom Kangaroos in North America; Trouble on Brean Beach; Return to the Canyon; plus news, reviews and more

The Journal of the Centre for Fortean Zoology

Animals & Men
The Journal of the Centre for Fortean Zoology

THE GIANT TAILED TOAD OF DEVIL'S ARSE

Identifying felid prints; The sad story of the Blue Monkey; Hoaxes (three different ones); Warning signs on Wikipedia;A new twist on o.o.p butterflies; news, reviews and more...

Typeset by Jonathan Downes,
Cover and Layout by SPiderKaT for CFZ Communications
Using Microsoft Word 2000, Microsoft Publisher 2000, Adobe Photoshop CS.

First published in Great Britain by CFZ Press

**CFZ Press
Myrtle Cottage
Woolsery
Bideford
North Devon
EX39 5QR**

© CFZ MMXIX

978-1-909488-60-1

For
Carl and Richard

Dear friends,

Welcome to this long awaited and, sadly, considerably overdue, collection of issues of *Animals & Men*. The Centre for Fortean Zoology has been in existence since 1992, and we have been publishing material on cryptozoology and allied disciplines for nearly all of that time.

In the past (nearly) three decades, the world of publishing has changed enormously; sometimes for the better and sometimes (in my humble opinion, anyway) not. I have been publishing magazines since I was ten years old, and by the beginning of the 1990s, we had it got down to a reasonably fine art. The first issue of *Animals & Men* came out in April 1994, and – like the other things that we were publishing at the time – was created using an electric typewriter, Letraset for the headlines, and a lot of glue. I wrote this somewhere else, and it was reposted on Twitter, whereupon various people commented on that last statement, assuming that my colleagues were using the glue for nasal gratification, rather than for the purpose which its manufactures intended. This is a totally base libel. The glue in question was Pritt Stick, and I doubt – with the best will in the world – that anything pleasurable would happen if you were to shove that up your nose.

But I digress.

From issue two onwards, we put the magazine together on a computer, albeit an Amiga 1200 games machine, but as of issue 26, we were fully online and in possession of a PC with a whopping 1.2GB of storage.

The subsequent years saw the arrival POD publishing and e-publishing, and these two innovations turned our *modus operandi completely upside down and inside out.*

Whilst we were now in the position to manufacture our own paperback books rather than having them photocopied and ring bound by a succession of what Richard Freeman called 'photocopier monkeys', the advent of streaming culture for magazines meant that, just as was already the case for music and films, the consumer was less and less willing to actually pay his or her hard earned cash for a magazine that they would rather read online.

Although *Animals & Men* had been available for hard copy for twenty years by then, we made the decision to publish it for free as on online flip-book, as well as publishing it, perfect bound, in the more traditional format. This was a great success, but sales of the traditional format continued to fall and – eventually – it got to the stage that it was no longer even partially viable to publish individual issues in hard *copy*.

Although the business ethic of CFZ Press (if you can actually call it that) was never profit orientated, it had been designed around the more traditional publishing models, and as these speedily began to change, our profit margin (such as it was) vanished like a sandcastle at high tide. So, we had to drastically rethink what we were doing and how we were doing it. And so, reluctantly, the decision was made to eschew publishing individual issues in hard copy, and to publish omnibus collections in book form.

Here, a little later than we had hoped, is the first of these collections. I hope that you enjoy it, and find it interesting.

I am particularly proud of what I have achieved in the last twenty-five years of *Animals & Men.* It is not only – as far as I am aware – the longest standing cryptozoological publication in the English speaking world, but it espouses a model of cryptozoology that places it well within the remit of the natural sciences, rather than as some peculiar branch of paranormal research. This is something that I, and the other leading lights of the Centre for Fortean Zoology, feel is utterly important, because – in our opinion – the internecine squabbles which take place across the internet about such hot topics as 'whether bigfoot has a cloaking device' or whether 'alien big cats are actually extra-terrestrial in origin' are completely counter productive, and do nothing except to drag what little good name cryptozoology still has through the gutter. Cryptozoology is not, or at least should not be, the study of ghosts, phantoms, or semi-decomposed dead raccoons, and – from the beginning –

the CFZ has done its best to foster an environment where cryptids are seen as real animals, and studied on that basis.

But, of course, we are not the Centre for Mystery Zoology. We are the Centre for *Fortean* Zoology.

From the beginning, I have always been interested in the stranger and non-animate phenomena that pepper our cultural psyche. Indeed, much though I have spent the last twenty five years trying to impress upon people that cryptozoology should be seen as a very real branch of the natural sciences, the book for which I am best known is – of course – my exhaustive look at the Cornish Owlman. It is this dichotomy; Flesh and Blood Cryptid vs. Zooform Phenomena that has fuelled the way that the Centre for Fortean Zoology has thought and acted over the past quarter century. And, I believe, it is reflected successfully in the contents of this long overdue book.

Enjoy.

Hare bol,

Jon Downes
(Director, Centre for Fortean Zoology)
August 2019

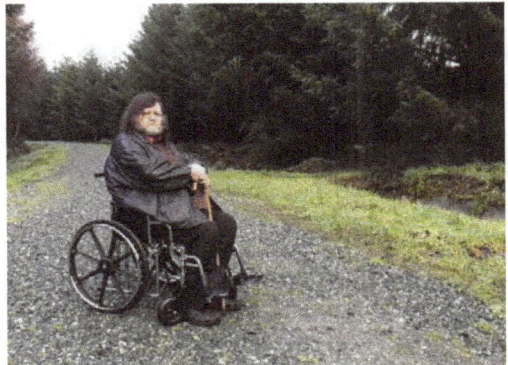

Animals & Men

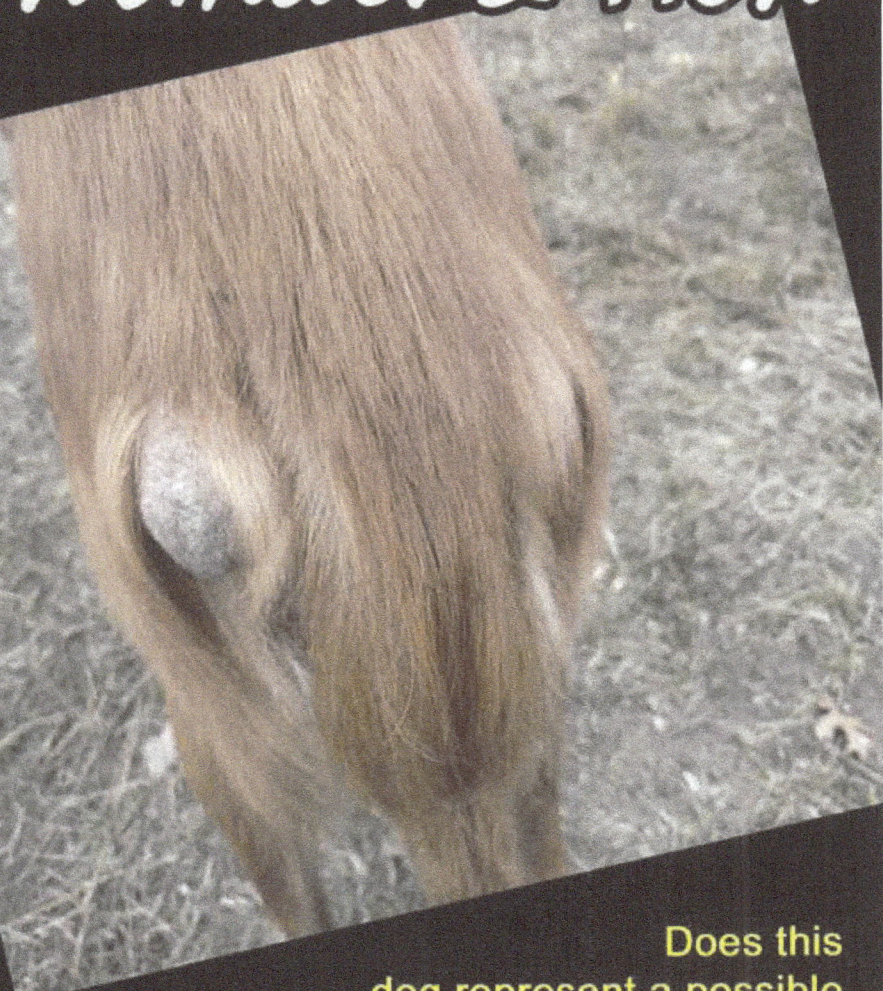

Does this dog represent a possible breakthrough in Blue Dog research?:

Plus: Tasmania 2016 expedition report; Mark A Hall Obituary; A Mystery Carcass in Djibouti; The Gorilla - Does it exist?; In Search of Bowheads and Narwhals; Weird Weekend 2016; News, Reviews

The Journal of the Centre for Fortean Zoology   #58/9

Contents

Typeset by Jonathan Downes,
Cover and Layout by SPiderKaT for CFZ Communications
Using Microsoft Word 2000, Microsoft Publisher 2000, Adobe Photoshop CS.
First published in Great Britain by CFZ Press

CFZ Press, Myrtle Cottage, Woolsery, Bideford, North Devon, EX39 5QR

© CFZ MMXV

ISBN: 978-1-909488-31-1

Faculty of the Centre for Fortean Zoology

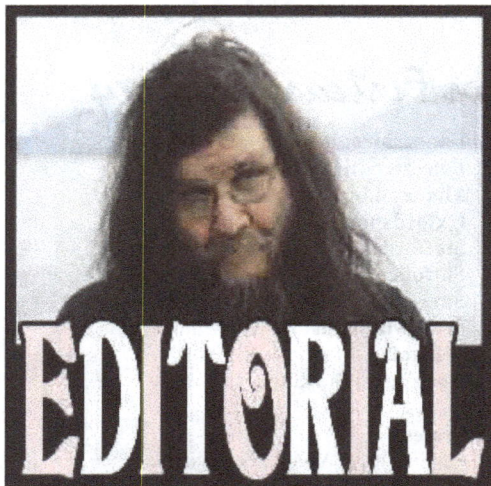

EDITORIAL

Dear friends,

Welcome to another issue of, what I sincerely believe to be, the worlds longest running cryptozoological magazine. It seemed such a frivolous conceit back in 1994 when I first

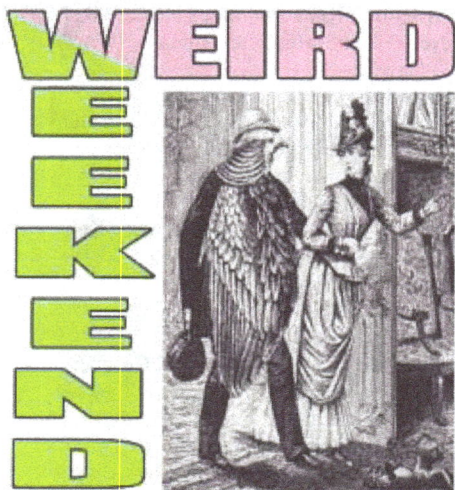

WEIRD WEEKEND

Une Semaine de Freaky

started the magazine, and I never really thought about it being still published 22 years later.

But, still being published it is, and I am faced with the very real possibility that if it is still being published in another 22 years time that I may not be there to se it.

I am not being melodramatic here, but I am not in good health and the way that my body has deteriorated over the past decade or so was the main reason that I made the decision that I did at this year's Weird Weekend.

Apparently it came with somewhat of a shock to some people who were there, which surprises me because I have written about it being a possibility several times on my Facebook page and on the CFZ blog. Those who wish to see my closing address can do so

The Great Days of Zoology are not done!

courtesy of Alan Dearling on YouTube. I began by jokingly referring to the fact that various members of my family have been (and still are) preachers of one kind or another over the years. So, I said that I was going to follow

I then played an excerpt from David Bowie live at the Hammersmith Odeon, London in July 1973 when he retired on stage. He said :

"Of all the shows on this tour, this particular show will remain with us the longest, because not only is it the last show of the tour, but it's the last show that we'll ever do."

I was surprised (and I will admit gratified) that although this wasn't greeted with the hysteria that greeted Bowie's announcement, there where audible gasps. Of course it wasn't the end of David Bowie's career, and it could be argued that even after his death in January this year his career is looking pretty damn healthy. But, it was the end of the career of his alter ego Ziggy Stardust. I explained that I have been running the Weird Weekend for 17 years now and that it is getting too much for me.

I can now reveal that the Small School in Hartland, who have played host to us over the past three years, are closing for good. This is something that I have been aware of for the last year, but was asked not to make public. So, I explained, I do not know whether there will be another Weird Weekend, but if there is, it won't be next August, it won't be in Hartland, and it won't be raising money for The Small School. And it probably won't be in the same

"Of all the shows on this tour, this particular show will remain with us the longest, because not only is it the last show of the tour, but it's the last show that we'll ever do."

in the family tradition and preach a sermon. And that I had chosen for the text of my sermon a quote from one of my own particular band of saints.

form as it has been.

Three lots of people have so far taken it upon themselves to announce that they will be taking over the event in Devon, and I have had to

The Weird Weekend (2000-2016)

After seventeen years and seventeen events, CFZ Director Jon Downes announced at the end of this year's event that the Weird Weekend is on hiatus. This does not mean that there will never be another event held in Devon, indeed there probably will be. But when, where, and how are - at the moment - in the lap of the Gods.

Despite rumours to the contrary, although we have authorised events in Cheshire and Denmark next year, we have not given our blessing to any other events, and when there is another Devon Weird Weekend it will be promoted by us!

Watch this space.

crack down upon this, because I have no intention of letting the event be taken over by any Tom, Dick, or Harriet, and at least one of the people who has decided to jump into my shoes before I have even decided whether I will or will not be doing another event, is one of the people least well equipped to take the event over that I could have thought of.

And, as I said on stage at the WW, I want to stress that whether or not I do another Weird Weekend, the CFZ will most definitely continue, at least while I am alive, and

hopefully for long afterwards. The CFZ has always gone through phases of its existence and we are merely entering another one.

Some of the young people whom I hoped would get more deeply involved in the running of the organisation have moved on to other things. This is perfectly within their rights, and I want to stress that this is perfectly OK with me, and that the comments on Facebook a few days before the WW castigating "young people whom the CFZ has nurtured" were not authorised by me, and do not reflect my views.

But this does lead me onto the next phase of my editorial.

Facebook and other social media have made the 'global village' a real thing to idealistic young hippies like I was four decades ago. But it also brings with it an unforeseen downside. Now anybody can share their opinion with the world whether or not the world wants to listen. And this can have very adverse affect on an organisation like this. I usually try to ignore what is written about us on Facebook, only stepping in to correct misapprehensions like that outlined above.

Just as we were going to press we had a very nasty shock. As a result of equipment failure, probably after a power outage, Graham found our entire colony of Rio Cauca caecilians dead in their tank. It was a horrible shock for us all, because we had nurtured them for over ten years, and it was - as far as I am aware - the only breeding

weird weekend 2016

colony in the UK. We are all shocked and terribly upset. It was particularly nasty for Graham who was "very fond of the little wriggly things".

Next year sees our quarter century in existence, our Silver Anniversary if you will. Whether or not there is a Weird Weekend in Devon next year, we shall continue to do the job for which I founded this organisation back in 1992. That is not to bicker and get upset about what people say on Facebook, but to approach the solving of zoological mysteries in a reasoned and scientific manner, and to push back the boundaries of human knowledge, and in doing so change the world… just a little bit.

Regular CFZ watchers will be aware that, although we have tried very hard, we are not exactly keeping to schedule at the moment. This is not, as one unpleasant correspondent claimed last year, because we spend all our time having long coffee breaks, but because, at the moment we have a number of family issues and health problems with which we have to deal. I, for example, have just been told that I have serious ulcerations on both feet. My left foot is particularly bad, and I have been told that if we cannot get this under control I may have to face severe consequences, up to and including the possibility of amputation. I have tried to make light of this by making silly jokes about being 'legless' but the truth is that I am terrified by the prospect, and the thought of the possible necessary treatment is eating into my time to an alarming extent. However, I am happy to say that issue 4 of the *Journal of Cryptozoology* is now for sale. However, in order to give us a little much needed breathing space, this *Animals & Men* is a double issue, and the last one of 2016.

Hare Bol.
Jon

Obituary

MARK A HALL
(1946-2016)

My extremely close friend, colleague, co-author, and fellow Midwesterner, Mark A. Hall, died on the morning of Wednesday, September 28, 2016. He was such a quiet force in the foundation thinking taking place in cryptozoology, it may be years before the younger researchers realize what a loss this is. But for me, it is immediate and is a powerful punch to the gut.

Mark's brother had passed away a few years

ago, and then his brother's widow, Shelia had become a lifeline for Mark in recent years. Shelia Hall was able to be with Mark last night, when he talked for over an hour to her, saying again that he was happy his research was in good hands with the International Cryptozoology Museum. She was able to be with him today, and shared that Mark passed away very peacefully.

Mark had been challenged by various forms of cancer for years, winning sometimes, and more recently, losing the battle despite conversations and communications that he was only suffering temporary set-backs. He clearly wanted to write more, think more, and discover more answers to the mysteries that interested him. I am overwhelmed with grief at the loss of such an important contemporary of mine.

Mark was born on June 14, 1946, in Minneapolis, Minnesota, and turned 70 years old on his last birthday.

It always seemed fitting that Hall entered the world on Flag Day, as he was interested in politics and how the nation should be run. A Fortean, cryptozoologist, author, and theorist, Hall was raised in the heartland of America, in or near Bloomington, Minnesota (except for one temporary trial attempt at living in North Carolina). Mark served in Army intelligence in West Berlin, in the midst of the Cold War, as a Russian linguist, translating messages gleamed from behind the Berlin Wall. Besides being an editor at an archeological society in Minnesota after his military service, Hall went on to work in human relations in various branches of the federal government, mostly for the Department of Agriculture and the Customs Service, while in his home state. He was an old-fashioned patriot who allowed himself to question the scientific establishment with every breath he took, and with each stroke of his computer keys.

Mark A. Hall was intrigued by nature's anomalies for most of his life. For almost sixty years, he was actively pursuing historical records and eyewitness testimony concerning cryptozoological phenomena. He traveled extensively throughout the Americas – and the world.

I first corresponded with Mark when Ivan T. Sanderson introduced us through letters late in the 1960s. Mark was living in Minnesota, and I was living in Illinois. Before long, we would visit each other and engage in long conversations about unknown hominoids, cryptozoology, and our latest theories. On his first visit with me at my home in Decatur, Illinois, after my wife-at-the-time excused herself, Mark and I stayed up into the early morning hours talking and talking, and completely forgot about sleeping. I always found Mark one of the most intellectual thinkers in the field, and we fed off of the others' ideas and the challenges to our thoughts, via visits, letters, and phone conversations, long before emails and the Internet.

Mark actively researched mystery cats, hairy hominoids, surviving anthropoids, ancient civilizations, and hundreds of other topics. He had extensive files and a large library. One concrete action he took before he died was to pass along over ten large boxes of his binders of original research materials to the International Cryptozoology Museum. He wanted his research to find a home where future generations could learn more from his work. I wished to help his archives to be part of his living legacy. His files arrived this month, days before today came.

Mark A. Hall was a Fortean, and wrote Fortean articles, for a time a column for Fortean Times. For years, he edited and published the journal Wonders, devoted mostly to cryptozoology and to other Forteana that interested him. As a book author, some of his works, such as an early book on Thunderbirds, Natural Mysteries, The Yeti, Bigfoot & True Giants, and Living Fossils, were self-published. In recent years, he saw a couple of his books more formally published. They cover unique subjects, such as Thunderbirds: America's Living Legends of Giant Birds (now a rare paperback book, more available in hardcover). Mark always had other books waiting in the wings.

As a bold theorist, Hall wrote that North America is home not only to the Bigfoot of the Patterson-Gimlin footage, but also to drastically different primates such as the True Giant (probably Gigantopithecus, he thought) and the Taller-hominid (which he saw as survivors of the recorded fossil known as Homo gardarensis).

Hall and I, on behalf of Ivan T. Sanderson, kept track of the Minnesota Iceman throughout the Midwest, in 1968-1969, and I interviewed Mark about his involvement with the Minnesota Iceman for my just-completed "Afterword" in Neanderthal: The Strange Saga of the Minnesota Iceman (Anomalist Books, 2016). Mark also appeared on Unsolved Mysteries detailing his Minnesota Iceman entanglement. The case was featured as a part of the September 25, 1994 episode of Unsolved Mysteries on NBC, and then in repeats on the Lifetime Channel for decades.

Hall served as the director of the Sanderson-founded Society for the Investigation of the Unexplained (SITU) in the early 1970s, right before Sanderson's death in 1973.

Besides Mark's theories that challenged many even within cryptozoology, he had an incredible memory for details, and his insights always impressed me. In 1999, I coined the name "Marked Hominids" to honor Mark A. Hall. He had invented the concept of the "Taller-hominid," which was not well-understood within the field. I renamed Mark's "Taller-hominids" with the term "Marked Hominids," to describe these specific type of Bigfoot-like creatures seen from Siberia to the Eastern USA and elsewhere, who are often noted as aggressive, piebald ("marked"), and not actually the Neo-Giants, the Bigfoot/Sasquatch so well-known from the Pacific Northwest. The fact Mark had first described them as separate from Bigfoot seemed a natural to thus be christened "Marked Hominids."

Darren Naish takes on Mark's Bighoot theory and some of mine, in what Naish called "speculative creature building" a la' Ivan T. Sanderson, at TetZooCon 2015. Even in ridicule, a few people listened to Mark's thoughts. Mark felt, in time, his theories would make sense, based on future discoveries. Hall was not evangelical about his material, and thought the data could speak for itself, eventually.

Other than Mark's Unsolved Mysteries appearance and the 2005 conference appearance, Mark tended to avoid public appearances and the media. However, he did do a few overnight radio interviews.

Mark A. Hall and I appeared together on Coast to Coast A.M. with George Noory about "Mothman & Thunderbirds," on Tuesday, November 15, 2005. I presented an overview of the Mothman case, and we talked of how the creature has similarities to a giant owl, and may have used air turbulence from cars to facilitate its flight. Mark joined the show to share material on Thunderbirds, which reportedly have a wingspan of 18 to 20 ft– twice the size of any known birds. Hall recounted the 1977 Lawndale, Illinois case, where a young boy was picked up and carried briefly by one of the giant birds. Though unharmed, the boy's hair turned gray after the incident. Hall said there were similar instances in the 1800s.

Mark Hall did appear alone, with Ian Punnett, on Coast to Coast A.M., on June 30, 2006, on the show "Giant Owls & Thunderbirds." Hall talked about reported sightings of giant owls and other mysterious behemothic birds. Hall said American Indian legends, as well as modern accounts speak about such birds, particulary in the West Virginian Appalachian Mountains. Some of these "great owls" are reportedly man-sized with 10-ft wingspans, Hall explained, and were probably the inspiration behind the Mothman stories of the 1960s. Hall coined the term "Bighoot" to describe the true origins of the Mothman sightings, which he related to giant owls.

Then George Noory invited him back on July 20, 2006, to talk about Thunderbirds and other mysterious creatures. One cryptid Hall discussed was the "Lizardman"– a 7 ft. tall scaly creature that was reportedly seen in South Carolina in 1988. Hall related the Lizardman to folkloric tales of mermen and mermaids, who were said to inhabit the water yet be able to remove their fish parts like scuba gear when they came on land and sometimes mated with humans. On the topic of unknown hominids, in the Yukon Territory and Alaska, Hall talked of what he felt were modern day reports of Neanderthals seen wearing crude clothing and carrying axes.

Hall proposed a new technique and coined the name for a modern form of studying cryptids, which he termed "telebiology," years ago. He formalized his thoughts about the concept, in print, almost two decades ago. Speaking of his suggested method, specifically to study unknown primates, Hall wrote:

"With temporary captives, we should do the best we can with them and then set them free. The results will be genuine knowledge in the records we will then have, and we will have invested in the future of a new relationship with our primate relatives. This approach is part of what I have called 'telebiology,' a means by which we can begin to study the cryptids that have been the object of cryptozoology. If we make the effort to study animals at a distance, using our brains and technology, we can succeed where others have failed in the past. If we can accept that starting to study a species with a dead animal can be difficult, then we can put that goal at the end of the process instead of making it a requirement to do anything at all". ~ Mark A. Hall, from The Yeti, Bigfoot & True Giants, 1997: 110.

From 1992, for almost a decade, Mark published a journal called WONDERS that covered many mysteries in the fields of cryptozoology, anthropology, and the uncharted world of forteana.

The essence of Hall's earlier, now unavailable self-published Lizardmen serves as the foundation for a book-in-the-works that I hope some someday will be published, Merbeings: The True History of Mermaids and Lizardmen by Mark A. Hall, from the International Cryptozoology Museum. Mark became one of the first honorary members of the International Cryptozoology Society earlier in 2016.

Thanks for hanging in there so long, Mark, and for sharing so much with all your friends, family, and other folks down through the years. I will miss him, deeply. This world is much less interesting without Mark A. Hall.

LOREN COLEMAN (First appeared http://www.cryptozoonews.com/hall-obit/ used with thanks)

Newsfile

Black is Black

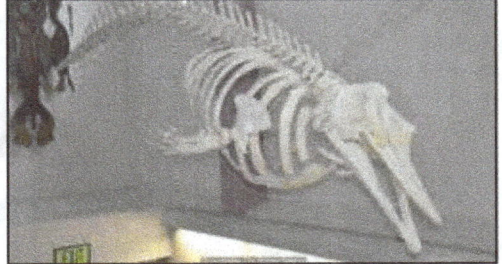

An international team of scientists who searched out specimens from museums and remote Arctic islands has identified a rare new species of beaked whale that ranges from northern Japan across the Pacific Ocean to Alaska's Aleutian Islands.

Japanese whalers call the enigmatic black whales "karasu," the Japanese word for raven. The new species is darker in colour and about two-thirds the size of the more common Baird's beaked whale, but so scarce that even whalers rarely see them.

A DNA analysis of 178 beaked whales from

around the Pacific Rim found eight known examples of the new species, the scientists reported today in the journal *Marine Mammal Science*. The eight included specimens from the Smithsonian Institution and Los Angeles County Museum of Natural History, a skeleton on display in an Alaska high school, and another that puzzled researchers trying to identify it when it washed up on an island in the Bering Sea.

"The challenge in documenting the species was simply locating enough specimens to provide convincing evidence," said Phillip Morin, a research molecular biologist at NOAA Fisheries' Southwest Fisheries Science Center, and lead author of the new study. "Clearly this species is very rare, and reminds us how much we have to learn about the ocean and even some of its largest inhabitants."

SOURCES:
- http://www.bbc.co.uk/earth/story/20160729-the-new-whale-nobody-has-seen-alive
- https://www.sciencedaily.com/releases/2016/07/160726123757.htm

Spot that Neck

It is a famous, gentle giant of the African savannah, but the giraffe's genetics have just revealed that there is not one species, but four. Giraffes have previously been recognised to be a single species divided into several sub-species.

But this latest study of their DNA suggests that four groups of giraffes have not cross-bred and exchanged genetic material for millions of years. This is a clear indication that they have evolved into distinct species. The study published in the journal *Current Biology* has rewritten the biology of Earth's

tallest mammal. The scientists say their findings could inform the conservation efforts for all four species of giraffe.

SOURCE: http://www.bbc.co.uk/news/science-environment-37311716

More Mouse Lemurs

For the first time in over 50 years, a northern quoll has been captured in Karlamilyi – Western Australia's largest and most remote national park. It's the first sighting from traditional owners since the 1960s and the first European record of the species in the area. The historic find comes after a joint survey effort between Kanyirninpa Jukurrpa (KJ) Punmu and Parnngurr ranger teams and

Department of Parks and Wildlife (DPAW) scientists earlier this month.

Located in the Pilbara, in the heart of Martu country, the park was once home to quolls – called wiminyji by the traditional owners – however they had not been seen since traditional life was disrupted around 50 years ago. "They recall seeing wiminyji throughout the Karlamilyi region when they were living a traditional life in the area, which was as recent as the early 1960s," said Gareth Catt, fire management officer for KJ. Read on...

SOURCE: http://www.australiangeographic.com.au/news/2016/09/first-live-quoll-found-in-was-largest-national-park-since-the-1960s

Hassan / Hemiptera

As many as 24 assassin bugs new to science were discovered and described by Dr. Guanyang Zhang and his colleagues. In their article, published in the open access *Biodiversity Data Journal*, they describe the new insects along with treating another 47 assassin bugs in the same genus. To do this, the scientists examined more than 10,000

specimens, including *Zelus truxali* (above) coming from both museum collections and newly undertaken field trips.

Assassin bugs are insects that prey upon other small creatures, an intriguing behaviour that gives the common name of their group. There are some 7000 described species of assassin bugs, but new species are still being discovered and described every year. The new species described by scientists Drs Guanyang Zhang, University of California, Riverside, and Arizona State University, Elwood R. Hart, Iowa State University, and Christiane Weirauch, University of California, Riverside, belong to the assassin bug genus Zelus. Linnaeus, the Swedish scientist, who established the universally used Linnean classification system, described the first species (*Zelus longipes*) of Zelus in 1767. Back then, he placed it in the genus Cimex, from where it was subsequently moved to Zelus.

All of Zhang & Hart's new species are from the Americas. Mexico, Panama, Peru, Colombia and Brazil are some of the top countries harbouring new species.

SOURCE:http://phys.org/news/2016-07-assassins-radar-species-assassin-bugs.html#jCp

A Modern Prometheus

Prometheus, the mythological Greek heroic deity, has been given a namesake in a new species of tiny rain frog, discovered in southwestern Ecuador. The name was chosen by the international team of scientists, led by Dr Paul Szekely, Ovidius University, Constanta, Romania, in acknowledgement of the Prometeo program, funded by the Ecuadorian government. The description of this new species (*Pristimantis prometeii*) is the result of the cooperation between three Romanian Prometeo investigators affiliated with the Universidad Tecnica Particular de Loja and Universidad Nacional de Loja, and two Ecuadorian specialists from Pontificia Universidad Catolica del Ecuador. The full study is available from the open access ZooKeys.

SOURCE: https://www.sciencedaily.com/releases/2016/07/160721143446.htm

Toads to Newcastle

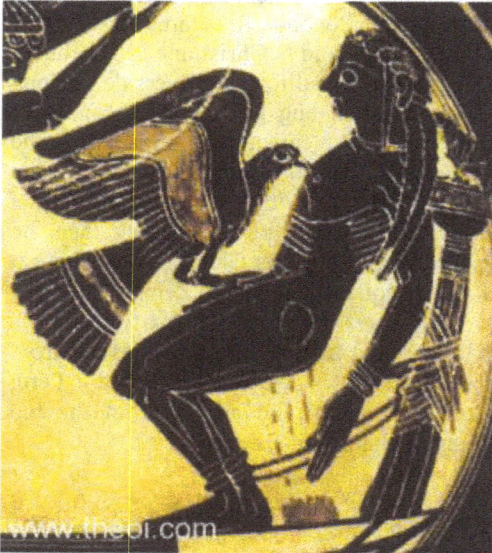

A biologist from Newcastle, NSW, has made a remarkable discovery – a species of frog previously unknown to science, found in Port Stephens, just 10km from Newcastle Airport. Dr Simon Clulow, a biologist at the University of Newcastle, said it was the frog's unique marbled black and white underbelly that led him to realise he had found something special. "The distinctive marble pattern on the frog's belly, along with other features makes it quite different to any other frog species in this part of the world and led us to believe straight away that we had found a new species – it was an incredible moment," said Simon, who first spotted the new frog at a sand swamp at Oyster Cove.

Another Compendium of Batrachia

Discoveries of new vertebrate species are rare, but particularly so within developed areas like Newcastle, which is the second most populated area in NSW after Sydney.

SOURCE: http://www.australiangeographic.com.au/news/2016/11/new-frog-species-found-in-newcastle

Brazilian Batrachian

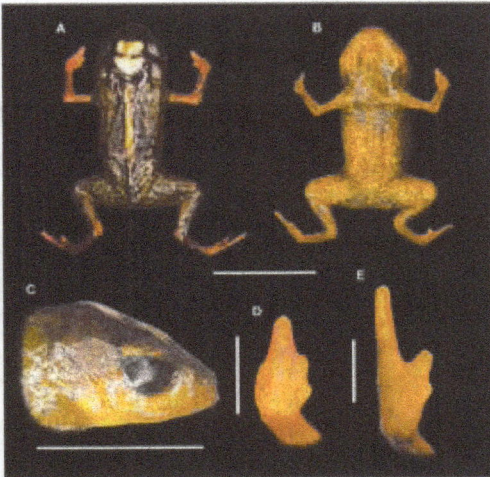

A new species of *Brachycephalus* (Anura: Brachycephalidae) is described from the Atlantic Forest of northeastern state of Santa Catarina, southern Brazil. Nine specimens (eight adults and a juvenile) were collected from the leaf litter of montane forests 790–835 m above sea level (a.s.l.). The new species is a member of the *pernix* group by its bufoniform shape and the absence of dermal co-ossification and is distinguished from all its congeners by a combination of its general coloration (dorsal region of head, dorsum, legs, arms, and flanks light, brownish green to dark,

olive green, with darker region in the middle of the dorsum and a white line along the vertebral column in most specimens) and by its smooth dorsum. The geographical distribution of the new species is highly reduced (extent of occurrence estimated as 25.04 ha, or possibly 34.37 ha). In addition, its habitat has experienced some level of degradation, raising concerns about the future conservation of the species. Preliminary density estimates suggest one calling individual every 3–4 m^2 at 815–835 m a.s.l. and every 100 m^2 at 790 m a.s.l. Together with the recently described *B. boticario* and *B. fuscolineatus*, the new species is among the southernmost species of *Brachycephalus* known to date.

SOURCE: https://peerj.com/articles/2629/

Sleeping Beauty

On the foothills of the Andes in central Peru, a brilliantly coloured frog lives out a fractured fairy tale. Once upon a time—specifically, one evening in November 2014—biologist Germán Chávez heard a call echo through the highest-altitude forests of Tingo María National Park. Chávez didn't recognize the call, so he went out to find the source of the distinctive chirps. Up in a tree, he found a little brown frog less

Another Compendium of Batrachia

than an inch (2.5 centimetres) long—and quickly realized that like the frogs in stories, this one was more than meets the eye. "We could see the bright red legs, and that was a surprise," says Chávez, a biologist with Peru's Center of Ornithology and Biodiversity. "We have never seen a frog like that."

"The exact function [of the pigment] for this species? We don't know," says co-discoverer Alessandro Catenazzi of Southern Illinois University. "But we start with putting a name on it." After two years of analysis, Chávez and Catenazzi have confirmed that the scarlet-shanked frog represents a species new to science. Its name, *Pristimantis pulchridormientes,* or the sleeping beauty rain frog, is a nod to the mountain range where the frog was found, which locals describe as resembling a sleeping reclined woman.

SOURCE: http://news.nationalgeographic.com/2016/08/species-frogs-andes-peru-rainforests-sleeping-beauty/

The Grand Old Frog of York

A STRIKING NEW frog species named the Cape York graceful tree frog (*Litoria bella*) has been discovered by scientists in Far North Queensland. Closely resembling its

southern relative the dainty tree frog (*Litoria gracilenta*), the Cape York graceful tree frog had previously avoided detection.

Despite the physical similarities, the Cape York graceful tree frog differs from its close relative through morphology, genetics and call. The newly-discovered species is more closely related to a New Guinea species, *Litoria auae*, than to the dainty tree frog it's mistaken for.

SOURCE: http://www.australiangeographic.com.au/news/2016/09/new-frog-species-north-queensland

Littering in Sarawak

The Gunung Mulu National Park in northern Sarawak is considered to be one of the most amphibian-rich areas on the island of Borneo, with about 90 species recorded within the park (about half the amphibian diversity for the entire island). This includes four species of dwarf litter frog, Leptobrachella spp., more than has been recorded in any similarly sized area, including one that has never been recorded outside the park. Recent studies have revealed that the distribution of dwarf litter

Another Compendium of Batrachia

frogs is strongly segregated by altitude, leading to the possibility that this number may not reflect the full diversity of these frogs in the mountainous Gunung Mulu.

In a paper published in the *Raffles Bulletin of Zoology* on 5 August 2016, Koshiro Eto, Masafumi Matsui and Kanto Nishikawa of the Graduate School of Human and Environmental Studies at Kyoto University describe a fifth species of dwarf litter frog from the Gunung Mulu National Park.

The new species is named *Leptobrachella itiokai*, in honour of Takao Itioka, an entomologist at Kyoto University and a core member of the biological researcher consortium in Sarawak. It was discovered at an altitude of 1445 m, within the known range of *Leptobrachella brevicrus*, by the call of the males, which is different in the two species. Only adult males of the species have been observed, females and tadpoles being unknown (though it is possible that tadpoles previously recorded in the area and assumed to belong to *Leptobrachella brevicrus* are in fact *Leptobrachella itiokai*.

SOURCE: http://sciencythoughts.blogspot.co.uk/2016/08/leptobrachella-itiokai-new-species-of.html

Dainty Dendrobates

A new species of diurnal frog of the genus Aromobates is described from the Sierra de Perijá in the Andes of western Venezuela. The new species is the first dendrobatid reported from this mountain range, though many other congeners are known from the Cordillera de Mérida, also in the Venezuelan Andes. It can be readily distinguished from all congeners by the unique combination of the following characters: dorsal skin granulate, paired and protuberant dorsal digital scutes, finger I shorter than finger II, fringes absent on fingers I and IV, present and conspicuous on all toes, toe webbing basal, dorsolateral stripe present, oblique lateral stripe diffuse, ventrolateral stripe absent. With this new species the number of Aromobates species from Venezuela increases to 13.

SOURCE: https://www.researchgate.net/publication/263654988_A_new_frog_of_the_genus_Aromobates_Anura_Dendrobatidae_from_Sierra_de_Perija_Venezuela

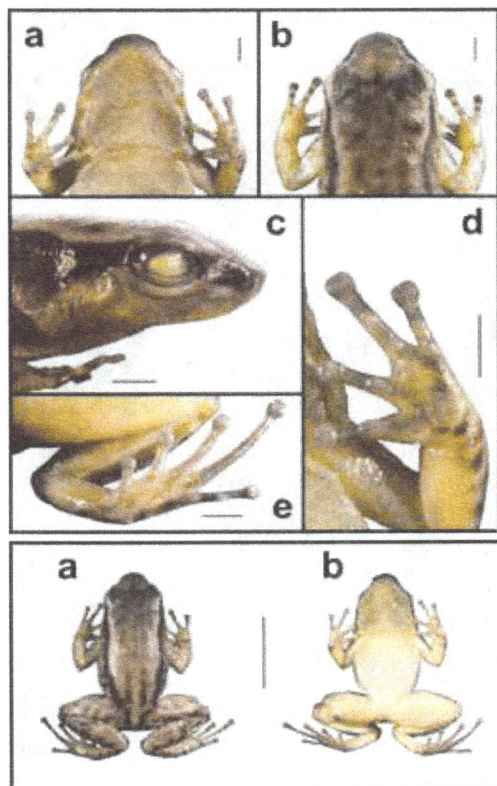

Another Compendium of Batrachia

Another Anole

A University of Toronto-led team has reported the discovery of a new lizard in the middle of the most-visited island in the Caribbean, strengthening a long-held theory that communities of lizards can evolve almost identically on separate islands. The chameleon-like lizard -- a Greater Antillean anole dubbed *Anolis landestoyi* for the naturalist who first spotted and photographed it -- is one of the first new anole species found in the Dominican Republic in decades.

"As soon as I saw the pictures, I thought, 'I need to buy a plane ticket,'" says Luke Mahler of U of T's Department of Ecology & Evolutionary Biology and lead author of an article on the discovery published online in *The American Naturalist*.

"Our immediate thought was that this looks like something that's supposed to be in Cuba, not in Hispaniola -- the island that Haiti and the Dominican Republic share," says Mahler. "We haven't really seen any completely new species here since the early 1980s."

SOURCE: https://www.sciencedaily.com/releases/2016/06/160617140558.htm

Costa Rica is the pits

An international team of scientists has solved a case of mistaken identity and discovered a new species of venomous snake.

The newly discovered Talamancan palm-pitviper is a striking green-and-black snake living in some of the most remote regions of Costa Rica. The colouring is a characteristic it shares with its close relative the black-speckled palm-pitviper. In fact, these two species look so similar that the Talamancan palm-pitviper went unrecognized for more than 100 years. It is a case of cryptic speciation, where two species look almost identical, but are genetically different.

"It's a really interesting phenomenon," said University of Central Florida biologist and professor Christopher Parkinson who led the team that made the discovery. "It shows some of the complexities we deal with when cataloging biodiversity and underscores the importance of maintaining natural-history collections. Discovering this species would not have been possible without the specimens housed in natural-history museums."

SOURCE: https://www.sciencedaily.com/releases/2016/07/160718142202.htm

Scorpio Rising

A riotously colourful new species of scorpionfish has been found deep in the Caribbean near Curaçao.

The fish is orange-red, with splashes of yellow and pink decorating its fins and face. Its scientific name is *Scorpaenodes barrybrowni*, after nature photographer Barry Brown, who works with the Smithsonian Institution mission that discovered the deep-sea-living fish.

S. barrybrowni is a denizen of the rocky seafloor and underwater cliffs, spending its time between about 310 and 525 feet (95 to 160 meters) down. Researchers discovered the new species during the Deep Reef Observation Project (DROP), a Smithsonian Institution mission to explore reefs deeper than scuba divers can go. Researchers used a manned submersible, Curasub, to collect samples of scorpionfish from near the island of Curaçao and discovered that several were of a species never seen before.

SOURCE: http:// www.livescience.com/55538-starburst-scorpionfish-discovered.html

The eagle-eyed amongst you will have noticed a minor change in these pages. I suggested to Karl Shuker that he follow my lead and use Tiny URL in the JoC but he pointed out that by doing so one reduces the value of the references for researchers. So, point taken, and although they do not look as neat, Tiny URL has been banished back to non referential use.

Don't tread on an ant (he's done nothing to you)

A research group in Hong Kong has described and named a new species of ant from Hong Kong, *Paratopula bauhinia*, or the rare "Golden Tree Ant." The Insect Biogeography and Biodiversity research group led by Dr Benoit Guénard at the School of Biological Sciences, the University of Hong Kong (HKU) has recently described and named a new species of ant from Hong Kong, *Paratopula bauhinia*, or the rare "Golden Tree Ant" in *Asian Myrmecology*, a peer-reviewed, yearly journal dedicated to the study of Asian ants.

A newly described species is a species previously unknown to science. The person who describes it has the right to name it. The new ant species discovered by the research team represents the 22nd ant species described from Hong Kong since 1858. The last one was in 2000. Descriptions of earlier species had to be dated back to 1928.

While some might think that new species are only discovered in deep pristine forests, this new ant species was found just a few hundred meters from HKU campus on the foothill of Lung Fu Shan Country Park during a night field course. The unusually large size of the ant (about 7mm long) and its golden appearance piqued the curiosity of Ms Ying Luo, a research assistant of the School of Biological Sciences, to collect it for further detailed inspection.

Back at the laboratory, she and Dr Guénard realised that this ant was quite special, not only did the specimen represent the first record of the ant genus Paratopula for Hong Kong and southern China, it also represented a new species for science.

Despite intensive collection efforts since its original collection, it has only been found at Lung Fu Shan Country Park and so the ant can be considered as endemic to Hong Kong Island.

This new species is described in *Asian Myrmecology* by Ms Luo and Dr Guénard, along with the first description of the queen of another arboreal species, *Rotastruma stenoceps*.

SOURCE: https://www.sciencedaily.com/releases/2016/08/160803075752.htm

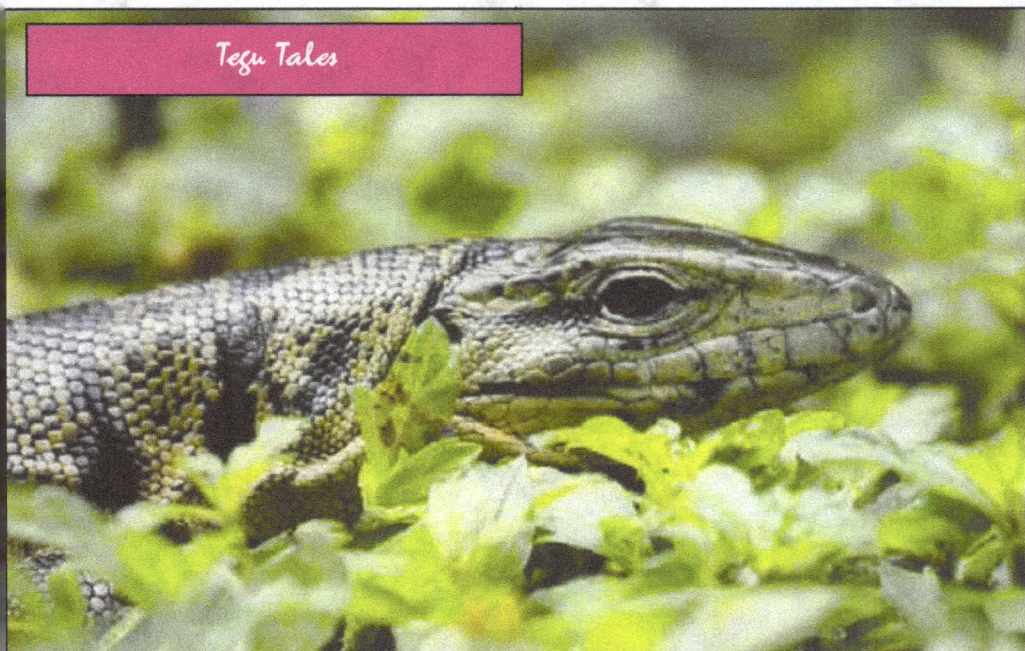

Tegu Tales

The golden tegu lizard, previously thought to be a single species, may actually comprise four distinct clades, including three new cryptic species, according to a study published August 3, 2016 in the open-access journal PLOS ONE by John Murphy from the Field Museum of Natural History, USA and colleagues.

Tegus are among the largest Neotropical lizards, and while some species occur only in Brazil, *Tupinambis teguixin* inhabits much of northern South America. Commonly known as the golden tegu, *T. teguixin* is also sometimes called the "black and white" tegu and can be confused with the closely related species, *Salvator merianae*. To help resolve the systematics and nomenclature of this species, the authors examined museum samples of golden tegus for genetic and morphological differences across its geographical distribution. The authors noted subtle differences in leg scale morphology, as well as the shape of eye and lip areas, and identified substantial genetic divergence across the tegus large range. The authors split the species currently recognized as *T. teguixin* into four morphologically distinct but geographically overlapping species, including three new cryptic species -- *T. cryptus, T. cuzccensis,* and *T. zuliensis* -- that look similar to the human eye but are genetically distinct. The authors suggest that further research in northeastern South America might identify additional species within the T. teguixin group, which would aid in planning for tegu conservation. "We demonstrate for the first time that two lineages of the Golden Tegu, *Tupinambis teguixin,* are living side by side at multiple locations in South America, and that *T. teguixin* is composed of at least four distinct species," said John Murphy.

SOURCE: https://www.sciencedaily.com/releases/2016/08/160803151104.htm

From the Hills of Montezuma

Mexican scientists recently described a new and strikingly colored species of earth snake from the mountains of Puebla and Veracruz in east-central Mexico. These burrowing reptiles are seldom encountered and, consequently, have been poorly studied. Furthermore, several species have restricted distribution, making them particularly vulnerable to extinction. The description of the new species was published in the open access journal *ZooKeys*. Looking to shed light on the evolutionary history and diversity of earth snakes, Luis Canseco-Márquez and Adrián Nieto-Montes de Oca, Universidad Nacional Autónoma de México, started to collect samples of these rarely seen creatures. A sample from east-central Mexico exhibited a unique set of traits among earth snakes, one of them showing striking orange and black banding pattern. They realized that these snakes represented a new species and proceeded to describe it in collaboration with scientists Carlos J. Pavón-Vázquez, Universidad Nacional Autónoma de México, and Marco A. López-Luna, Universidad Juárez Autónoma de Tabasco. The specimens of the new species were deposited in the herpetological collections of the Museo de Zoología "Alfonso L. Herrera" and the Instituto Tecnológico Superior de Zongolica.

SOURCE: https://www.sciencedaily.com/ releases/2016/08/160811131646.htm

The Crab with the Golden Claws

Ever since I used to catch them and keep them as pets when I was a boy in Hong Kong, the colourful freshwater crabs of South China have been of intense interest to me. They are in great demand for the pet trade pet keepers across the world. To answer the demand, fishermen are

busy collecting and trading with the crustaceans, often not knowing what exactly they have handed over to their client. Luckily for science and nature alike, however, such 'stock' sometimes ends up in the hands of scientists, who recognise their peculiarities and readily dig into them to make the next amazing discovery. Such is the case of three researchers from University of New South Wales, Australia, The Australian Museum, Sun Yat-sen University, China, and National Chung Hsing University, Taiwan, who have found a new species and even a new genus of freshwater crab, and now have it published in the open access journal *ZooKeys*.

Knowing about the growing demand for eye-catching freshwater crabs from southern China, the authors took a look at the ornamental fish market to eventually identify an individual with unusually structured male gonopod, which in crustaceans is a swimming appendage modified to serve as a reproductive organ. Having their interest drawn by the peculiar crab, lead author Chao Huang managed to persuade the fish dealer to let them survey the collection site located in northern Guangdong, southern China. Despite superficial resemblance to an already existing freshwater crab genus, at second glance, the crab turned out to be quite distinct thanks to a unique set of features including the carapace, the gonopod and the relatively long and slender legs. Once the molecular analyses' results were also in, the authors had enough evidence to assign the freshwater crab as a species and even a genus new to science. Being a primarily aquatic species, the new crab prefers the pools of limestone hillstreams, therefore its name *Yuebeipotamon calciatile*, where calciatile means 'living on limestone'.

SOURCE: https://www.sciencedaily.com/releases/2016/09/160907125305.htm

Thylacine

Thylacine Videos

One of the most interesting pieces of video to come our way during the last few months, came in mid-September: a video of an unknown creature that had been filmed in the suburbs of the Adelaide hills started doing the rounds. Leaving aside the idea of one of the rarest animals in the world being filmed in broad daylight in a built up area, Corinna makes a very good point :

"What I don't understand though: At the beginning when it is just a film being taken on camera (why was it filming anyway over the road at someone's bins and trees?) it only seems to catch the middle and tail end - no head - before the camera points up into the trees. Although when slowed down and zoomed etc the head is suddenly visible?"

She also notes:

"It is not a cat I don't think, because if you freeze frame it at the correct moment you can see it has a long head"

There is also the important question of why somebody is carefully filming the wheelie bins outside a suburban house. These are the sort of comments that do lead the 'I want to believe' brigade to accuse us of being negative naysayers, but that is the way the cookie crumbles. Nobody would be happier than us to be confronted with conclusive evidence that the Thylacine still exists but, I am afraid, that – at the moment – we cannot

accept that this video is it. Whilst the second video is more impressive, it is hard not to agree with Tina Indyka on the Facebook page for the Thylacine Awareness Group who writes: "Interesting footage -I thought the hopping gait looked more like a creature with an injury and though it had a straight tail the head looked more foxlike to me - I guess everyone sees something different and we all want to believe"

And the third video, which surfaced as we were going to press, is superficially even more impressive, but we are withholding judgement until more information surfaces.

Man Beasts (BHM)

Any Port in a Storm

It's a great image, but I don't believe a word of it. In #56 (p.28) of this journal we covered a report that had appeared in the British press suggesting a yeti had been photographed in the Spanish ski resort at Formigal. There are a number of reports of man beasts from the Pyrenees, although I have my doubts whether these things could actually be flesh and blood, animals. I am even less convinced by the photograph which appears to show a bloke in a monkey suit. A few days afterwards, a website called *The Lad's Bible* printed a story confirming that the yeti sighting was nothing more than a marketing ploy by the company that owns and runs the ski resort.

And as I wrote at the time: "Now, why are we not surprised?"

The trope of European manbeasts seems to be travelling far and wide at the moment, and these images are by far the most entertaining, if not impressive. The big problem is - for me at least - is that none of the news stories trumpeting the arrival of this video can tell us anything about its provenance.

We don't even know where in Portugal it was taken, although one claim is that it is on a small piece of desert in Madeira which is - apparently - the only desert in Portugal. However, despite the fact that I have a mate who is a senior editor there, *The Daily Express* and *Daily Mirror* are not noted for

their veracity, and one needs to consider the possibility that they were just being lazy journalists and that they may just be talking about the Desertas Islands which is a small archipelago in Madeiran waters which isn't a desert at all. However, at least they didn't claim that if the UK had not voted for Brexit, a whole horde of these creatures, probably both single parent families and supporters of ISIS, would come rushing through the Channel Tunnel demanding Social Security benefits.

I have been studying the fauna of Madeira and other Macaronesian islands in the Atlantic, and am fairly au fait with their biology.

And although their fauna includes some interesting anomalies, the idea of there being a man-sized mammal lurking in even the wildest parts of any of these islands is an absolutely ridiculous notion.

Bizarrely, however, the "thing" (as I am sure that Ivan T Sanderson would have described it) depicted in the stills which I have shown in these pages, looks remarkably like the other "thing" that I and several other members of a CFZ expeditionary force saw in Bolam Woods, Northumberland during January 2003.

What this means, or doesn't mean, I don't like to hazard a guess.

SOURCE:

- http://www.dailymail.co.uk/news/article-3743586/Is-mythical-chupacabra-Mysterious-man-like-creature-filmed-roaming-Portuguese-desert.html
- http://www.express.co.uk/news/weird/701032/Man-like-creature-bigfoot

Bonny Dundee

And now for some Bigfoot sightings from the UK. As I have already alluded to, I saw such a "thing" in January 2003, and - to use the current vernacular - in the interests of full disclosure, and also because I do not want to appear to be one of those authors who take every opportunity to sell their books, here is an excerpt from my 2004 autobiography, *Monster Hunter*:

"What I saw was an incredibly flat and angular figure, which appeared to be two-dimensional. It is here that I have greatest difficulty in describing my encounter because what I saw were so far out of the normal run of human experiences that there simply are not adequate words to describe it. I will do my best, but even now - over a year after the expedition was done and dusted - I still find it almost impossible.

As far as its shape is concerned, the nearest analogue that I have been able to come up with is the angular metallic running man which can be found in certain levels of the computer game Doom II. However, this "thing" (as Ivan T Sanderson would undoubtedly have called it), was a matt-black. It was so black in fact that it was a quasi-human piece of nothingness, which had somehow become projected upon the Northumbrian landscape. It moved far too sharply and far too fast to be a living thing - at least in the ways that we know it. Again I have to resort to another unwieldy and fairly inadequate analogue. It was as if somebody had filmed this humanoid shaped piece of nothingness, and then projected it back on to the landscape using the fast-forward facility on the video recorder. The whole experience lasted only a second, but it has been with me ever since.

When the expedition returned on Monday, we conducted experiments to find out exactly how far away the creature - if it was a creature - was from the excited onlookers. Using Richard as a model, I made a fairly accurate estimate that the creature had been a hundred and 34 feet away from him at the time of his sighting. I also estimated or that the creature had run along a distance of between 12 and 18 feet".

So I am not someone who is going to dismiss such sightings out of hand, but as I said then, and have said at various times in the intervening twelve years, the idea that there are flesh and blood higher primates lurking undiscovered in the British countryside is completely ridiculous, and so - whatever these "things" are, they cannot be explained using a purely zoological model, and the following accounts should be read with this in mind.

Mark Luke, 46, from Edinburgh broke his silence for the first time after struggling to come to terms with what happened for over a decade. He decided to speak out after stumbling across an online article from last year on *The Courier* website while researching his own alleged encounter. Mr Luke, who is a carer, said he spotted the mythical beast just off the Tay Road Bridge in 2005 whilst travelling to Dundee to fill in for nightshift security cover. His sighting happened in the same year that a former civil servant from Fife reported seeing a creature at the Five Roads Roundabout when he was driving home from work.

Just like Mr Luke, the civil servant had also told nobody at the time and only spoke publicly about what happened some 10 years later. Mr Luke said: "I had a similar encounter driving back to Edinburgh from nightshift security in Dundee, around 8am, on the A92 south in the same rough area. I can remember a forest and I looked to my left to see a tall dark shape standing 20 feet away in the trees."

SOURCE: https://www.thecourier.co.uk/ news/local/fife/66569/i-was-quite-scared- and-confused-edinburgh-man-believes-he- spotted-bigfoot-in-fife-countryside/

The Welsh Wonder

In September, a Bigfoot researcher claimed to have recorded one in Wales. Paul Seaburn wrote it up for Mysterious Universe.

"To my astonishment, I saw what I can only describe as a gorilla-like figure, slightly blended into the dark bush, but still having a clear form, just like a gorilla or Bigfoot creature might have."

Jason Parsons claims he saw this creature in the woods on Caerphilly Mountain near Cardiff while looking for Bigfoot markers as part of his hobby which, not surprisingly, is Bigfoot hunting.

"I had previously visited Caerphilly Mountain a few days before, and noticed plenty of the typical wooden 'X' type structures believed to be Bigfoot boundary markers. Upon my third visit, while walking up a trail, whether human or Bigfoot is unclear, I noticed an oddly broken twig. I proceeded to film this stick, whilst talking and explaining to the camera what I saw. I did not notice at the time, but later on, while viewing the footage for editing, I noticed movement in the background."

Seaburn was not very impressed with this video, and I am afraid to say that I have to agree with him. But we remain open minded enough to wait and see what happens next.

SOURCE: http:// mysteriousuniverse.org/2016/09/first-video- ever-of-a-possible-bigfoot-in-the-uk/

Surabaya Johnny

Half a world away we have another, equally blurry, video which does appear to show a black, human-like figure walking across and beneath a waterfall. What no-one else has commented upon, to the best of my knowledge, is that it seems to be carrying something round and shiny.

Over, once again, to Paul Seaburn, who is truly adept at pointing out the flaws in such footage:

"The appearance of the creature in the video is brief and so are the details on the sighting, which is disappointing for a channel that calls itself Bigfoot TV. The tall (some viewers estimate it to be 8 ft or 2.4 m high) creature walks in front of the waterfall and disappears. While it doesn't seem to be bathing, there are comments that it's glistening as if it just stepped out of a shower. Some commenters see a pail or water bucket in its hand, others spot two or three more creatures in the shadows. The quality of the video seems good but doesn't hold up well under magnification. Indonesia has plenty of waterfalls so trying to pinpoint the location is impossible without more information."

Indonesia is an archipelagic country extending 5,120 kilometres (3,181 mi) from east to west and 1,760 kilometres (1,094 mi) from north to south. According to a geospatial survey conducted between 2007

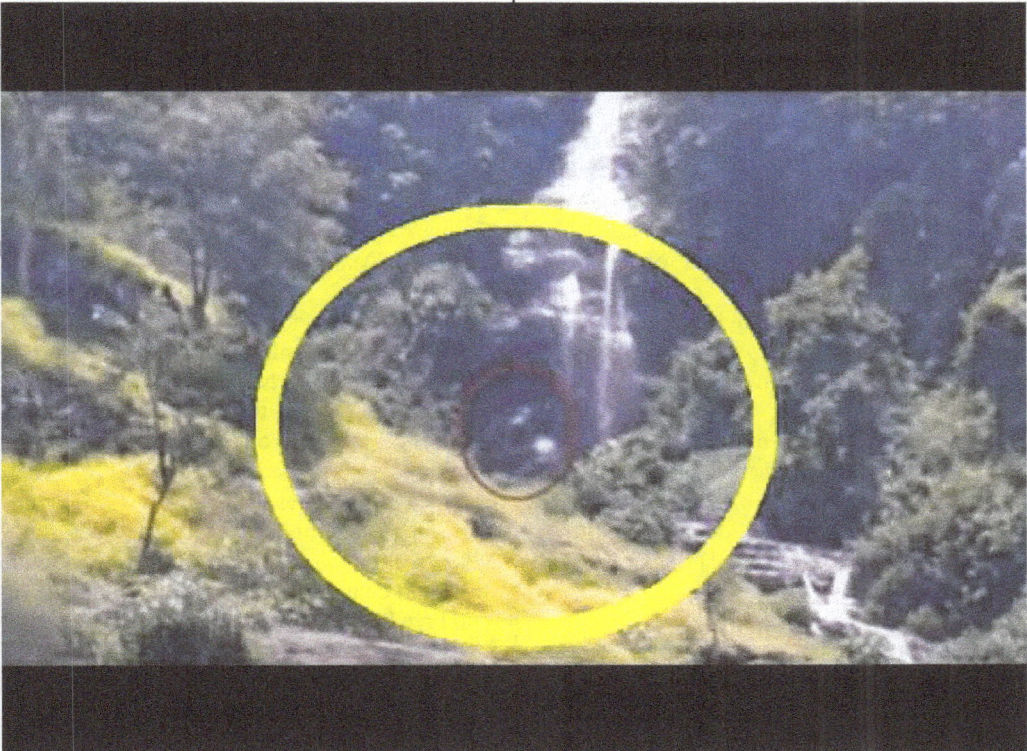

and 2010 by National Coordinating Agency for Survey and Mapping (Bakosurtanal), Indonesia has 13,466 islands. However, according to earlier survey conducted in 2002 by National Institute of Aeronautics and Space (LAPAN), the Indonesian archipelago has 18,307 islands. According to the CIA factbook, there are 17,508 islands.

The discrepancy of the numbers of islands in Indonesia was caused by the earlier survey includes "tidal islands"; sandy cays and rocky reefs that are appeared during low tide and submerged during high tide. There are 8,844 islands have been named according to estimates made by the government of Indonesia, with 922 of those permanently inhabited.

With a country where people can't even settle on the number of actual islands there are, unsupported video evidence with no site reference is worse than useless.

The video certainly shows a humanlike figure walking along and under the waterfall, and - to my mind at least - there is no evidence to suggest that this human-like figure isn't a human resident of the area going about his or her ablutions, and (assuming that the shiny object being carried is a net of some sort) possibly doing a spot of fishing at the same time.

Now I am going to sit back and wait for the death threats, and accusations that I am a "debunker" (certainly true) who doesn't believe in anything to do with Bigfoot (certainly untrue) to arrive. Pah!

SOURCE: http://mysteriousuniverse.org/2016/10/bigfoot-spotted-showering-in-an-indonesian-waterfall/

The Eagle has Landed

An eagle cam appears to have captured an ape-like creature walking around on the ground below, prompting speculation that it could be the elusive creature.

The shadowy figure was filmed near the Platte River State Fish Hatchery in Beulah, Michigan, back in May, *The Detroit Free Press* reported. The camera had been set up to monitor a resident eagle family's progress by the state's Department of Natural Resources and outdoor website CarbonTV. It wasn't until recently, however, that a CarbonTV editor stumbled across the video's clip on a Bigfoot sighting website and highlighted it on theirs, said CarbonTV executive director of content and marketing Daniel Seliger.

The footage has since gone viral with an edited version of it magnifying the creature as it walks, seemingly on two legs, around trees and other brush. Though there appear to be some believers – at least on social media — not everyone is convinced it's the real deal. That includes Bigfoot researcher Jeffrey Meldrum of Idaho State University. "It is an interesting video, but remains just that, given the lack of scale and detail in the image of the figure," he told Free Press after reviewing the video.

SOURCE: http://www.huffingtonpost.com/entry/possible-bigfoot-seen-in-eagle-

Mystery Cats

European Round Up

● Costa del Sol, Spain

A VIDEO published on social media on Tuesday, July 26, appears to show a large, cat-like beast prowling around in scrubland close to Estepona, on the Spanish Costa del Sol. Commenters speculated as to the identity of the mystery animal, which is said to weigh approximately 30-35 kilos and was only on display for a few seconds before disappearing into a thicket of trees.

SOURCE: http:// www.euroweeklynews.com/3.0.15/news/on-euro-weekly-news/costa-del-sol-malaga/140115-is-a-big-cat-on-the-loose-in-estepona

● Gloucestershire, UK

Peter Lennon and his wife stopped in the middle of the M50 in Gloucestershire when the spotted the mysterious large cat cross the carriageway at night on Sunday 24th July.

The pair were driving home to Redditch from Cardiff when an animal "the size of a mountain lion" appeared from the bushes at the side of the road. Peter said the "dark brown or black" beast stood next to the couple's van for a moment, before disappearing into the night. Now Peter and his wife are convinced they have seen a panther or a lynx.

SOURCE: http://www.dailystar.co.uk/news/ latest-news/533318/black-panther-britain-couple-sighting-spotted-mysterious-beast-motorway

● Yorkshire, UK

Royal Mail worker Phil Chapman couldn't believe his eyes when he spotted this cat like creature roaming around the countryside near Wakefield. Mr Chapman, 48, was staying at the *Redbeck Motel* in Crofton when he captured the animal, believed to be a big cat, on camera, as it was prowling through nearby fields. Sadly, no matter what I do with Photoshop, all I can see is an amorphous blob that could be anything. Sad but true.

SOURCE: http://www.wakefieldexpress.co.uk/ news/big-cat-captured-on-video-panther-like-beast-spotted-in-countryside-1-8091437

● Hertfordshire, UK

Iain MacDonald told the *Herts Advertiser*: "I had a rather interesting experience whilst out jogging on Sunday (9) October, to the north of Harpenden, when a big cat – I am pretty sure it was a panther – crossed the road directly in front of me."

It is the latest in a string of recent alleged

sightings of a panther as *Herts Advertiser* allegedly learned three weeks before, one is believed to have been spotted in nearby Hitchin, only 15 miles away. Iain explained that he "was jogging, out towards Luton Airport, in Chiltern Green. I was running along a quiet country road, with hardly any cars, where there was woodland on each side and this animal loped across the road, about 50 metres in front of me. It didn't look at me or anything, but it was going from one wooded area to another, on the other side."

Iain said he was "so surprised that I didn't register it straight away, and it took a while to work out what I saw. It was a very large cat, about four to five feet long, and had a very long tail. It was walking low across the road – I sped up a little bit!" Iain said that he often jogs in the area, "but I haven't seen something like that before. I didn't take a photo, as I was jogging". He added: "It was black and I'm 100 per cent sure that it was a panther. Everyone I have told thinks I must have been drinking, but I'm absolutely convinced about what I saw. Some people have had big cats as pets, and it was substantially bigger than a [domestic] cat!"

SOURCE: http://www.hertsad.co.uk/news/ panther_on_the_prowl_harpenden_man_spots_a_ big_cat_near_luton_airport_1_4734223

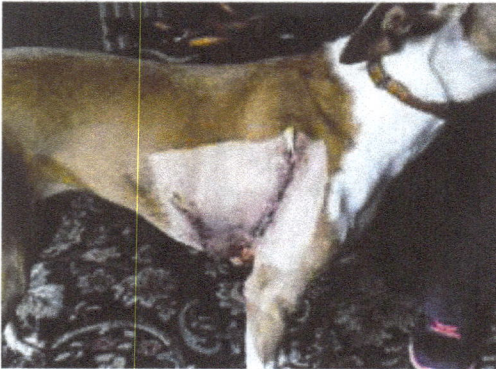

- *Buckinghamshire, UK*

The Sun carried the following story: "A DOG nearly died from horrific injuries inflicted in a

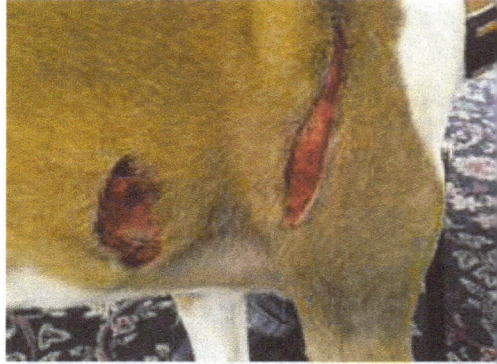

savage attack by 'the Beast of Bucks'. Rescue pooch Daisy was being taken for a walk by owner Carlos Romero, from High Wycombe, in woodland near local beauty spot Tom Burt's hill when she was attacked by a 'black panther'.

Terrified Carlos believes his beloved pet was pounced upon by a jungle lion as the area is a hotspot for sightings of big cats. In 2001, experts confirmed prints found on Wycombe Heights Golf Centre were those of a puma. Since then, locals have reported several sightings of a black big cat in the area, leading it to be nicknamed the "Beast of Bucks".

Carlos was walking Daisy, who is a mixed breed, in the woods when she went off to investigate something in the undergrowth. Minutes later she howled in pain and came out limping towards her owner with blood gushing from gaping wounds on the side of her body."

The article goes on, in dramatic prose, to proclaim that these big cats are a threat to children, and that: "If it is a panther, somebody needs to do something about it". From where we are sitting however, I see no evidence that a big cat had anything to do with the attack on poor Daisy, which could equally well have been done by a badger, a fox or another dog. But that doesn't sell newspapers.

SOURCE: https://www.thesun.co.uk/ news/1968116/dog-suffered-horror-wounds-after- being-savaged-by-a-puma-on-country-walk-

Aquatic Monsters

It ain't N

There are times that I rather like my life, and - when I am sitting down on a grey Tuesday afternoon, critiquing the forthcoming Pink Fairies album, and writing about recent sightings of anomalies at Loch Ness - this is one of them.

Sceptical student, Jolene Lin, was on a Loch Ness tourist boat, hunting for the lake monster. The 24-year-old noticed something moving in the waters and took a photo of what she described as a creature with a 'snake-like head'. She told the Sunday Post: 'About midway through the journey, I looked to the right of where I was standing and I saw a thin snake-like head emerging from the waters and moving a little before submerging back into the waters again after a few seconds.

'Thankfully I was constantly taking pictures while on the boat ride so I managed to capture what I saw then, and took two pictures of the encounter.' Jolene said she was a non-believer of Nessie, but after taking the pictures she is a converted believer.

SOURCE: http://metro.co.uk/2016/08/07/are-these-pics-proof-the-lock-ness-monster-is-not-a-myth-6053290/#ixzz4PQiAbFba

This second photograph is probably the best photo - technically, at least - ever taken of something strange in Loch Ness. But, as usual, when something seems too good to be true it usually is.

A whisky warehouse worker has excited Nessie hunters after photographing something mysterious in Loch Ness - but it appears to be just three seals. Ian Bremner, 58, was driving around the Highlands in search of red deer - but stumbled across what he claims could be the world's most famous monster swimming in calm waters.

His photographs show what appears to be a dark beast with a long winding body bobbing on the surface between the villages of Dores and Inverfarigaig. Some friends have said the beast's head could be a seal and his picture captures the extraordinary moment three of them were playing together in the water. But he said: 'I suppose it could be seals - but I'm not so sure.

The more I think about it, the more I think it could be Nessie.' But Mr Bremner said he is sure it is Nessie and shows a two-metre long silver creature swimming away from the lens with its head bobbing away and a tail flapping a metre away, preparing to swim further on.

He said: 'It's a part of the world that always makes you second guess what you're seeing. 'When you're up there you're constantly looking in the water to see if you can spot anything in there. 'This is the first time I've ever seen Nessie in the loch. It would be amazing if I was the first one to find her.

© Ian Bremner / SWNS.com

At the CFZ we think that the photograph does indeed show grey seals playing, but it is interesting to note that even two decades ago it was generally thought that these creatures did not visit the lake.

Gambolling seals and/or otters have often been cited as being responsible for lake monster sightings, but—to the best of our knowledge—this is the first time that they have been photographed *in flagrante delicto.*

SOURCE: http://www.dailymail.co.uk/news/article-3792704/Whisky-warehouse-worker-claims-convincing-evidence-Loch-Ness-monster-looks-JUST-like-three-seals-playing-water.html#ixzz4PQkItJpb

The next pictures are far less impressive unfortunately. They were taken in August by Ian Campbell, 56, who was on a bicycle ride with his son and a family friend when he spotted two big 'creatures' apparently swimming across the Loch together. The council regulatory officer, who says he is not a man 'given to flights of fancy', is convinced that the two 'monsters' he saw and photographed were both about 30ft in length. He said: 'At the time we saw it we had stopped for a rest and to admire the view. It seemed to appear suddenly from nowhere. 'I said to my son: 'What is that in the water?' He said to me that it looked like a big animal.

'I said 'I think you're right' and grabbed my camera phone to take a picture.'

I am afraid that at the CFZ we do not have access to anyone who can extrapolate anything from these blurry pictures. However, we would not wish for a moment to imply that Mr Campbell was not acting in good faith when he made these pictures public, or that they did not see something of interest in the cold Scottish water. The only thing we can extrapolate is that his 'phone does not take very good pictures.

SOURCE: http://www.dailymail.co.uk/news/article-3781412/Is-proof-Nessie-s-friend-Council-officer-snaps-photo-TWO-30-foot-monsters-swimming-Loch-Ness.html#ixzz4PQoeMKZE

Finally, also in August, holidaymaker Calley Tulleth, 28, took this photograph from her holiday home overlooking the water. She said her eyes 'popped out' of her head and she grabbed her camera. "I was having my lunch on the balcony looking over Loch Ness and all of a sudden I saw this blue thing swimming across. I wasn't scared but my eyes popped out of my face". "I quickly took my phone out and started snapping away – I tried to zoom in and get a better view but then it just disappeared. It looked so much like the Loch Ness monster."

SOURCES:

http://www.dailymail.co.uk/news/article-3781412/Is-proof-Nessie-s-friend-Council-officer-snaps-photo-TWO-30-foot-monsters-swimming-Loch-Ness.html#ixzz4PQoeMKZE

https://www.thesun.co.uk/news/1626664/walker-claims-to-have-pictured-the-loch-ness-monster-during-recent-trip/

© Calley Tulleth / SWNS.com

Siberian Khatru

Investigator Andrey Solovyev is spending all winter on the shores of Siberia's notorious Lake Labynkyr, hoping to garner definitive proof of its legendary monster. 'I am not claiming anything, but I think that perhaps it does live here,' Andrey told representatives of the Oymyakon branch of Russian Geographical Society who ventured here recently to check on his progress. 'And I've got a chance to check it out. I didn't meet the [Labynkyr] devil yet.'

Yet then he qualifies his statement, feeling that he may have glimpsed it. 'Two weeks ago when the ice was still not completely frozen some dark creature swam out of the lake - but I couldn't identify it,' he said. Nor did he manage to capture it on camera, although he is ready to do so if he gets a definite sighting.

Could it have been the Labynkyr Devil? 'It is quite possible, but I can't be certain,' Andrey said. 'Some strange things were happening here, like in September when I set very robust nets and they were torn to shreds, like I could never imagine. 'I saw huge - meters in diameter - holes on these nets. It definitely wasn't done by a fish, even a pike couldn't bite through this net.'

SOURCE: http://siberiantimes.com/other/others/features/f0268-new-signs-of-siberias-nessie-a-dark-creature-in-the-lake-and-broken-fishing-nets/

Faked Alaska

This, not very convincing image is from footage, shot in the River Chena by Craig McCaa and Ryan Delaney of the Alaskan Bureau of Land Management flashed across the internet as so much balderdash does. In a blogpost on Hallow'een I wrote:

"Today's strangeness begins with a not very convincing photograph from Alaska reporting to be that of a river monster. http://tinyurl.com/jos55rf

So, of course, those jolly nice fellows of Her Majesty's Press have been contacting me in droves for comments. Basically this is what I wrote in return:

"That video from Alaska looks very much like this video from Iceland a couple of years ago: http://tinyurl.com/zevqqlj

That turned out to me an inanimate construct of ice in flowing water and I strongly suggest that this is something similar. Although there are reports of mysterious aquatic creatures from various parts or North America, including Alaska, I am not aware of any from that particular river.

Sorry about that."

And guess what? I was completely right.

SOURCE: http://www.thenational.scot/news/alaskan-nessie-turns-out-to-be-a-piece-of-rope-tied-to-bridge.24304

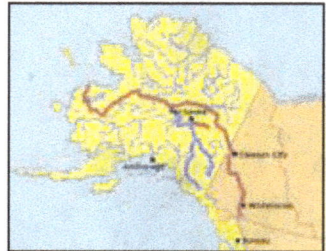

Newsfile Xtra

Some of this article appeared in the most recent issue of the CFZ Members Newsletter, but as there has been a significant development since, I am printing it here with amendments.

As regular readers of my cryptozoological writings will be aware, I have been studying the blue, hairless, dog-like creatures which have been reported from various parts of the Unites States ever since my then girlfriend and I went to Texas in November 2014 for an abortive audition for an American TV show called 'The Tracker'.

On the telephone the producer told me they where looking for somebody with a "rough and wild edge" and so I arrived wearing a leather jacket and having made no attempt to tame my long hair and beard. Despite having made a film about Sid Vicious, the director was appalled and it was obvious from the moment that we touched down at San Antonio airport that I wasn't going to get the job. Truthfully I didn't care much, being about as impressed with the director as he was with me. However, a couple of good things did come out of that trip (and I'm not talking about having charged an enormous room service bill to the TV company of course).

The first of these was that I was taken to a smallholding owned by Devin Macanally who had – earlier that year – shot a blue, hairless, dog-like creature that he had suspected of having eaten his prize mulberries (see the picture above). This whetted my interest, and I have been researching these animals ever since.

Six years later, Corinna and I returned to Texas as guests of Richie and Naomi West and we continued my researches. Richie and Naomi had discovered that the males of these blue dogs often had strange 'pads' of flesh on there haunches as can be seen in this specimen which was found as a road kill by Phyllis Canion.

More recently, Naomi West discovered the first photograph that we know of, of a female

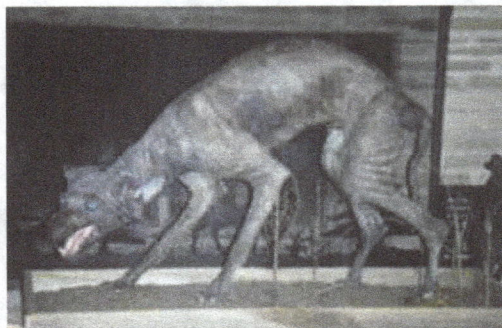

of one of these animals possessing these strange flesh pads which effectively disproved any suggestion that this was a form of sexual dimorphism. This can be found in issue #53 of *Animals & Men.* However, until now these weird excrescences have not been reported on any creature apart from the mysterious blue dogs. You will notice that I said until now. The second important thing to happen to me during my 2014 trip to Texas was that I met American cryptozoologist Ken Gerhard and it was one of his correspondents who posted the following photographs on Ken's Facebook page.

These pictures are of the hindquarters of a five year old golden retriever bitch called Lady.

She has not been neutered, and has no history of health problems. The only weird thing about her is that George Shank - a Michigan resident - whose niece owns Lady, believes that she may not be the pure bred retriever that she was advertised. She certainly looks too elongate to me. She has been examined by a vet who confirmed – just as Dr Canion said about her specimen - that these pads are pure flesh; not cancerous, and not glands of any sort. The veterinary surgeon had no idea what has caused this.

I am sending copies to my step-daughter Shoshannah who is a highly qualified vet (this means she is an animal doctor not that she fought in Vietnam) and to other people that may be able to solve the mystery. I have also set Richard Muirhead, who as many of you know is ridiculously good at hunting out arcane information, on to searching veterinary journals to see if he can come up with anything that might explain this. However, at the very least this means that we have a healthy animal with no signs of hair loss, which appears to be showing the same peculiar symptoms as so many of the Texas blue dogs.

Watch this space.

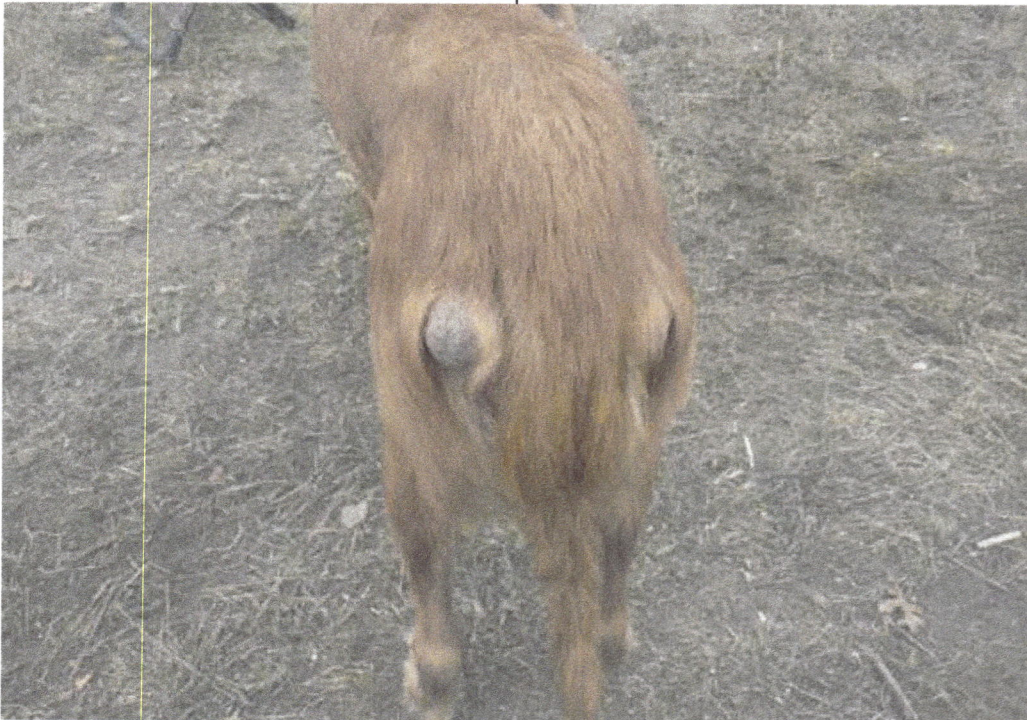

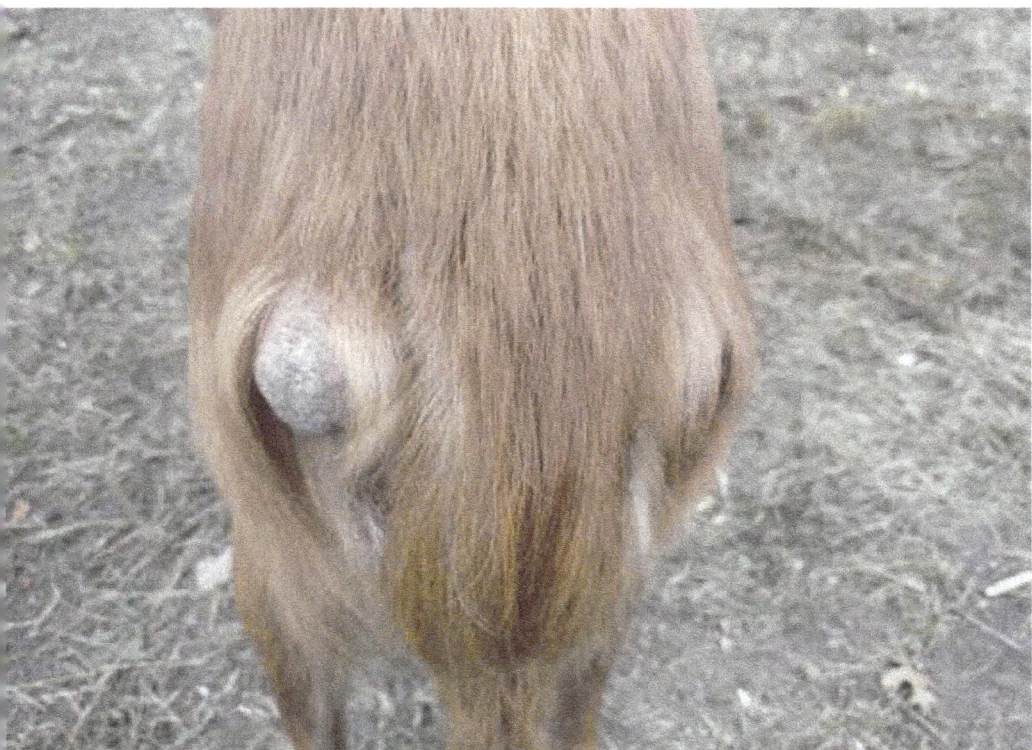

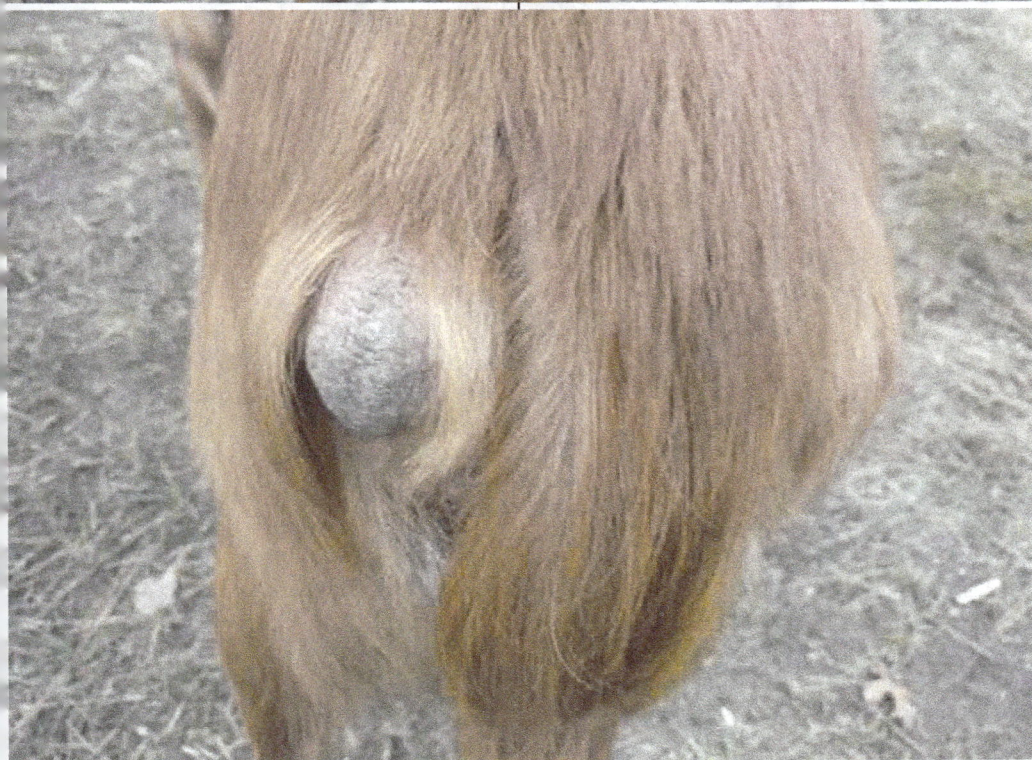

But then the plot thickens, as plots within the CFZ are wont to do. The very next day Naomi West sent us this picture of what appears to be a very dead and very flat canid. She wrote:

"Does this look like a blue dog to you? One of my students says it has absolutely no hair on it. He found it three days ago on a road in our town. He is going to go check and see if it is still there this evening. If he finds it, I've asked him to bag it (wearing gloves) and I will pick it up, but I have no plans after that. Richie and I really need to purchase a freezer for carcasses."

Sadly the carcass had disappeared by the time that the student returned for it, but Naomi and Richie have put the word out, and we are confident that we shall get an intact carcass eventually.

If YOU find one of these carcasses and do not have the facilities for storing it, can you please photograph every part of its body including close-ups of the skin. Please will you cut a one inch square out of a piece of card and use it as a transect, and count the number of visible hair follicles there are on the skin on different parts of the body. Please also take photographs of this for us as supporting evidence.

Newsfile Xtra

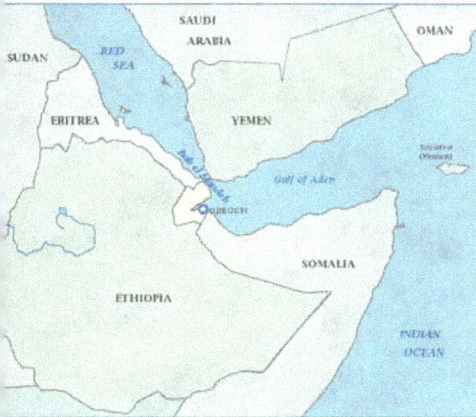

At the beginning of October I got the following e-mail. For reasons that shall be obvious to anyone reading this, I shall not identify the sender for security reasons. I may be an elderly hippy with a bad attitude but I come from a military family, and I do understand what is and what is not appropriate:

Hello Sir,

Here is those pictures from when I was stationed at Camp Lemmonier Djibouti Africa. That is not me in the photo but one of my navy buddies from the Sea Bees. I do have a video of it also if you think that might help. I know there is some more photos but I just cant find them now.

Well, thank you for taking a look for me. I am sure it is nothing new but with that area having such a large amount of migratory sea creatures who knows what could pass through those waters.

Lafe

My first reaction is that the pictures are of a dugong but Richard Freeman suggested that it is a pilot whale. On reflection I think that he is

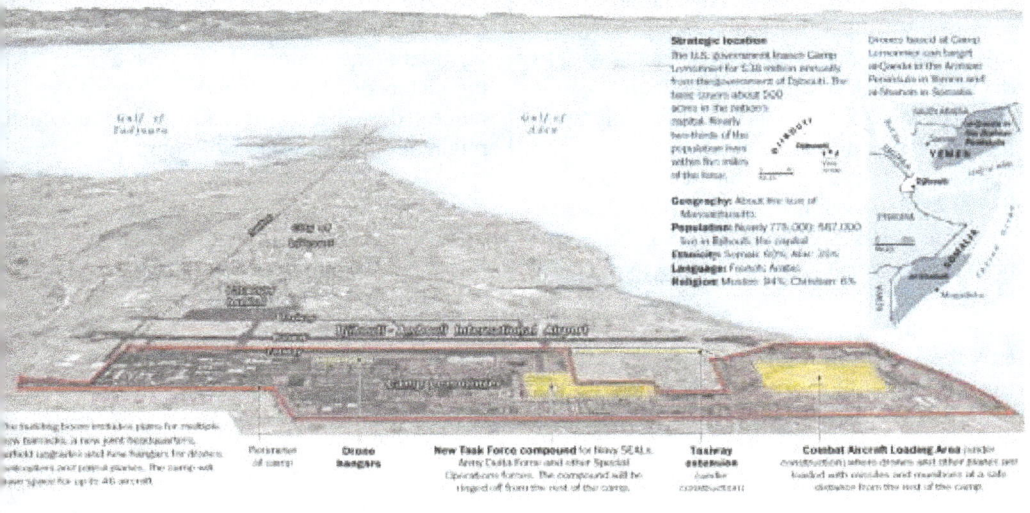

45

probably right, but I wrote back to Lafe asking permission to put it in this issue of A&M. He replied:

"As far as the pictures go, I myself thought maybe a young pilot whale. The mouth is what kind of threw me off. But I do not now what a pilot whale would look like if baked at 110 degrees for 3 days in the Djibouti sun. Maybe this? But since that region is a major migratory highway for many marine creatures, who knows what else could be passing through there? As far as publishing it in your newsletter please feel free to do so. I do not believe it to be the Patterson film nor the surgeons photo, but it would be neat to know. I am sure it is just some common creature. I am glad that you took the time to read my email. The only thing I ask is if you blur my shipmates face on the one photo. As he is back in the civilian world now and don't know how he would feel?

If you need anymore information or anything, please feel free to email me."

It is so refreshing to receive pictures from someone who is NOT claiming that they are priceless new evidence and wanting us to pay enormous sums for the privilege of looking at them let alone printing them. I have pixellated the pictures of Lafe's erstwhile shipmate as he requested, and also his uniform badge which might also be used to identify him.

Yes, it was me, so don't any of you conspiracy theorists get all paranoid on me. I am not going to let this become another Montauk Monster debacle. As we went to press Max Blake wrote: "Not sure exactly what it is, but it's certainly not a dugong". And so, at the risk of sounding like Esther Rantzen on *That's Life* over to you, readers...

watcher of the skies

CORINNA DOWNES

Before ornithologist John Young managed to photograph one in 2013, the night parrot (*Pezoporus occidentalis*) – described by Bush Heritage Australia's Jim Radford as a "dumpy budgerigar" or a "podgy, sort of smallish, green and yellow parrot", was thought to be extinct for more than 100 years.

However, the discovery of the species been recorded on an area of reclaimed pastoral lease now known as Pullen Pullen nature reserve in Diamantina National Park in central-west Queensland, expanding its known range and leading scientists to believe it may not be as rare as previously thought.

During October, another team of researchers from the Australian Wildlife Conservancy (AWC) and the Queensland Parks and Wildlife Service, led by Young, announced they had found what they believe to be a larger population of night parrots in the nearby national park.

Researchers made seven records of the bird this year: four sightings, three of which included nests with eggs, and three recorded calls. "My immediate reaction was excitement – this is great, there are more birds out there than we thought," Atticus Fleming, chief executive of AWC told Guardian Australia.

He continued, "But when you start to analyse it, the really significant thing about this is that these birds may be more common than we thought. That is something that we will be developing in the next few years as the study extends into other areas."

The Queensland government has declared that area a restricted access zone with hefty fines for unauthorised access to deter poachers or enthusiastic twitchers, and the same penalties apply for entering Pullen Pullen, which is owned and managed by Bush Heritage Australia.

BHA's head of science and research, Jim

NEW AND REDISCOVERED

Radford, listed poachers as one of the significant threats faced by the parrot, other threats including cattle, feral cats and potential habitat destruction from bushfires, which destroy the tall spinifex clumps where night parrots make their nests. Unfortunately, there are other dangers for the largely ground-dwelling parrot, and in April researchers from BHA discovered eggs in a night parrot nest after heavy rain, only to return later and find shell fragments containing traces of what proved to be the DNA of a brown snake. "Which is an interesting discovery in and of itself because we didn't realise that brown snakes would predate on eggs," Radford said.

The night parrot is one of just two fully nocturnal bird species in the world. The other is New Zealand's kakapo, famous for being the world's heaviest parrot. "I fully expect that they will be discovered in other places in Australia in time as well, because I don't think that this can be the only population," Radford continued.

SOURCE: http://tinyurl.com/zn83qxd

Karimui Owlet-nightjar - rediscovered after half a century

A Karimui owlet-nightjar was discovered live in August, after a gap of 52 years since its first discovery.

The Karimui owlet-nightjar (*Aegotheles affinis terborgi*) is only known from a single specimen taken near the Karimui Basin in the Eastern Highlands of Papua New Guinea in 1964, and since its discovery it had been placed under barred owlet-nightjar *A bennettii* based upon its close similarity to the two mainland races of *A bennettii*. It was, however, of a much

larger size, wing length being 154mm compared with *A bennetti bennettii* which has a 121-128 wing, as well as darker colouration. The *Handbook of the Birds of the World Alive* currently places it under Allied Owlet-nightjar *A affinis* and the recent *Birds of New Guinea - Distribution, Taxonomy and Systematics* places it under Barred *A bennettii*. Ashlet Banwell managed to take some photographs of it before it flew off, as well as some sound recordings at dusk the following evening.

SOURCES: http://tinyurl.com/hoz4txv
                http://tinyurl.com/jey9stn

Three new bird species discovered in Africa

During November, it was announced that a Texas A&M University team has discovered three never before documented bird species in Africa. Dr. Gary Voelker, professor and curator of birds in the department of wildlife and fisheries sciences at Texas A&M University, College Station, headed the recent discovery of a trio of similar African birds living near to each other, but that are different species which share no common genes. "The discovery of these three new species is a good example of the amount of potentially hidden diversity living in Afrotropical forests," Voelker said. "Our evidence runs directly counter to the belief of earlier research that said Afrotropical forests are static places where little evolutionary diversification has occurred. The areas were referred to as 'museums' of diversity, meaning they believed because many of the birds look similar across their ranges, then they probably were the same species. That's a point we are finding not to be true."

Voelker described the three new species as forest robins in the genus Stiphrornis; two from West Africa and one from the Congo Basin. "Each of the three represents a distinct lineage based on our genetic analysis," he said. "The three are further distinguished from already documented birds in the genus by clear differences in appearance such as wing and tail length and subtle differences in their plumage; one species has a distinctive song as well." He explained that as many of the birds within the species look a lot alike, there hasn't been much research done historically to find if they are in fact, different species. This lack of research also means there is much less well-preserved DNA available for genetic analysis and those specimens that are available are rather old. However, his team's recent collecting work in Africa has enabled them to add to the limited genetic material that several other museums have, which in turn, has allowed them to address the species diversification question.

"This overall lack of collecting bird specimens in Afrotropical lowland forests is likely inhibiting the discovery of any number of new species, though several apart from the three we studied have been

described in recent years," he said. "This suggests that a lack of sampling in the region, rather than a lack of obvious variation in the birds, is a key contributor to fully documenting avian biodiversity in lowland forests."

They are named Dahomey forest robin (*Stiphrornis dahomeyensis*) found in Benin and the central region of Ghana, the Ghana forest robin (*Stiphrornis inexpectatus*) collected from Brong-Ahafo and Central Regions of Ghana, and the Rudder's forest robin (*Stiphrornis rudderi*) discovered along the Congo River near Kisangani in the Democratic Republic of the Congo.

Journal Reference:
Gary Voelker, Michael Tobler, Heather L. Prestridge, Elza Duijm, Dick Groenenberg, Mark R. Hutchinson, Alyssa D. Martin, Aline Nieman, Cees S. Roselaar, Jerry W. Huntley. Three new species of Stiphrornis (Aves: Muscicapidae) from the Afro-tropics, with a molecular phylogenetic assessment of the genus. Systematics and Biodiversity, 2016; 1

DOI: 10.1080/14772000.2016.1226978
SOURCE: http //tinyurl.com/j2zhdg6

Galápagos faces first-ever bird extinction

Researchers from the California Academy of Sciences, San Francisco State University (SFSU), the University of New Mexico (UNM), and the San Francisco Bay Bird Observatory (SFBBO) used molecular data from samples of museum specimens to determine that two subspecies of vermilion flycatchers, both found only in the Galápagos, should be elevated from subspecies to full species status. One of these newly recognized species -- the characteristically smaller San Cristóbal Island vermilion flycatcher -- hasn't been seen since 1987 and is considered to be the first modern extinction of a Galápagos bird species. The findings were published online earlier this May in the journal *Molecular Phylogenetics and Evolution*.

"A species of bird that may be extinct in the Galápagos is a big deal," says Jack Dumbacher, co-author and Academy curator of ornithology and mammalogy. "This marks an important landmark for conservation in the Galápagos, and a call to arms to understand why these birds have declined."

"Access to museum collections such as the Academy's for pursuing these types of studies is invaluable," says Christopher Witt, study co-author and associate professor of biology at the University of New Mexico. "Preserved specimens can provide the crucial links needed to better understand how life on Earth evolved."

Two subspecies of the vermilion flycatcher, both found only in the Galápagos, were determined to be so genetically distinct that

the team elevated them to full species status: *Pyrocephalus nanus* (throughout most of the Galápagos) and *Pyrocephalus dubius* (only on the island of San Cristóbal). The latter -- significantly smaller and subtly different in color from the other species -- is commonly known as the San Cristóbal vermilion flycatcher and hasn't been seen since 1987.

"Wouldn't it be great if the San Cristóbal Vermilion Flycatcher weren't extinct? No one is looking, I'm pretty sure of that," says Alvaro Jaramillo, study co-author and biologist at the San Francisco Bay Bird Observatory. What exactly drove the San Cristóbal Vermilion Flycatcher to extinction remains unknown, but two invasive threats to the archipelago likely played a part: rats and parasitic flies (*Philornis downsi*).

"Sadly, we appear to have lost the San Cristóbal Vermilion Flycatcher," says Dumbacher, "but we hope that one positive outcome of this research is that we can redouble our efforts to understand its decline and highlight the plight of the remaining species before they follow the same fate."

SOURCE: http://tinyurl.com/zrzzwao

- **Booming call of the endangered Bittern once again being heard across Anglesey**

During October it was announced that, after 32 years, a bittern (*Botaurus stellaris*) had nested and successfully reared fledglings at RSPB Malltraeth Marsh in Wales during the summer. A type of heron, the bittern is a rare wading bird last known to have bred in Wales at Valley on Anglesey in 1984.

"It's been a long time coming!" said the reserve's site manager, Ian Hawkins. He explained that RSPB Malltraeth Marsh was established in 1994 with the goal of attracting this birds back to Wales.

Although local birdwatchers knew of the nest site, they agreed to keep away to avoid disturbing the hatchlings. The bittern is one of the UK's most threatened species of bird and is a shy, secretive bird most likely to be heard rather than seen. Bitterns were prized in medieval times as a banquet dish, and numbers fell sharply as the wetland habitats loved by the species disappeared.

SOURCE: http://tinyurl.com/hgxfqls

- **Rare bird returns to the West Country - after five thousand years**

After a gap of 5,000 years, a Dalmatian pelican (*Pelecanus crispus*) has been spotted in Britain. Remains have shown the birds used to breed in the Somerset area

around 5,000 years ago.

It was filmed at Braunton, North Devon during October, and is thought to have been blown off course towards the UK. It was first seen in Poland and Germany before arriving at Land's End earlier this year.

There are thought to be around 10,000-20,000 Dalmatian pelicans in the world with the largest colony at Lake Mikri Prespa in Greece.

SOURCE: http //tinyurl.com/zzht522

SIGHTINGS – VISUAL AND AURAL

- **Ultra-rare bird spotted over Newfoundland**

A common swift (*Apus apus*) was spotted over Cape Race lighthouse. There are no swifts native to Newfoundland and Labrador and the only species of swift to regularly visit eastern North America is the chimney swift, which does occasionally appear in spring and autumn. The common swift's summer breeding range runs from Spain and Ireland in the West across to China and Siberia in the East. They breed as far South as Northern Africa (in Morocco and Algeria), with a presence in the Middle East in Israel, Lebanon and Syria, the Near East across Turkey, and the whole of Europe as far North as Norway, Finland, and most of sub-Arctic Russia. They migrate to Africa by a variety of routes, ending up in Equatorial and Sub-Equatorial Africa, excluding the Cape. Common swifts do not breed on the Indian Subcontinent.

SOURCE: http://tinyurl.com/jmbovpn

- **Possible sighting of summer tanager in Pembroke, a very rare bird for this region**

On 10th September, a possible female summer tanager was spotted near Water Street and the entrance to Algonquin College parking lot in Pembroke, Ontario. The last time one of these birds was located in the area was about 10-15 years ago in Deep River.

The summer tanager (*Pirange rubra*) is the most common North American tanager in its range across eastern and southern United States. In Ontario, it is quite scarce and is therefore considered a rare bird. It is mainly found alone or in pairs near water and bottomland hardwood and riparian forests. While wasps and bees are the bird's favourite food, it also enjoys other insects, grubs, caterpillars, and fruit.

SOURCE: http://tinyurl.com/j45cwf4

- **'Purple Swamphen' at Minsmere**

On 31st July 2016, a 'Purple Swamphen' was found on the South Girder pool at Minsmere RSPB, Suffolk. There have been several previous occurrences of 'Purple Swamphen' in Britain, though these have generally related to Grey-headed Swamphen (*Porphyrio poliocephalus*) of the Middle East, Indian Subcontinent and South-East Asia, a species which is kept in captivity widely across Europe, including in Britain. A record from Cumbria in October 1997 was considered of indeterminate form, with mixed opinion over its origins, which were never definitively established. African Swamphen (*Porphyrio madagascariensis*), breeding widely across sub-Saharan Africa and north along the Nile Valley to Egypt (and also in Israel), is also kept in captivity in Europe and birds appearing to match this phenotype, with extensive green upperparts, have been recorded as presumed escapes across the region.

The Minsmere bird was a uniform blue

colour, with slightly paler blue head and entirely blue upperparts, ruling out both poliocephalus and madagascariensis and simultaneously identifying it as a Western Swamphen (*Porphyrio porphyrio*) of Iberia and the Western Mediterranean — the first confirmed record of this species in Britain.

SOURCE: http://tinyurl.com/zkbn7tf

• **Rare kaki make fleeting West Coast appearance after 37 years**

In October, two kaki (*Himantopus novaezelandiae*) – black stilts - were spotted on a West Coast dairy farm in New Zealand - the first such sighting in 37 years - and the Department of Conservation (DoC) is asking the public to keep an eye out for them. They are critically endangered, with less than 100 birds in the wild. Ranger Liz Brown said, "It's pretty neat to see where they ended up. After they disappeared, we thought the worst. We supplementary feed newly released birds for around six weeks post release, but happily, these two sussed out how to fend for themselves without our help."

Kaki have been intensively managed since 1981, when their population declined to a low of just 23 birds.

SOURCE: http://tinyurl.com/j7a9dr6

• **First colony of Red-billed Tropicbirds found on the Canary Islands**

Lanzarote Pelagics posted, on its Facebook page, the following information; "In 2014 and 2015 two red-billed tropicbirds (*Phaethon aethereus*) were present on Fuerteventura, Canary Islands, but there was no proof of breeding. In 2016 Tony Mulet located two birds, in the same place that they had been observed in both 2014 and 2015, and recorded that they were regularly attending the cliffs there.

With the species' northern advance, it may be expected that the number of sightings within British or Irish waters will increase. Indeed, there has been two records within just the last three years, both from Cornwall, off Pendeen on 18th August 2013 and off Porthgwarra on 28th August 2015.

SOURCE: http://tinyurl.com/zolkb2r

• **Siberian Accentor on Shetland**

During October a first for Britain was spotted at Scousburgh, Shetland. The Siberian accentor (*Prunella montanella*) is extremely rare in northern Europe. This bird breeds just inside the Western Palearctic, and is fairly common in the Polar Urals.

However, it remains a rare (albeit increasing) vagrant elsewhere in northern Europe. An influx of the species appears to be taking place this autumn, and it may well be that more British records follow — five have also been recorded in Sweden and four in Finland since 4th October.

SOURCE: http://tinyurl.com/zpxns9k

- **Glossy Ibis spotted in Fremington Pill, North Devon**

A rare visitor, the glossy ibis (*Plegadis falcinellus*) - also known as the European ibis - was seen at Fremington Pill in September. Devon Birds member Tim said, "They don't breed in Britain. They are from southern Spain and south Europe and they are known here as a vagrant species. They are sometimes spotted in south Devon, but more recently they have been coming further north."

"They have been coming to the UK from 1996, but in 2010 we had a flock of 20 in Devon. We will see more of them if the recent trends continue." The ibis is more commonly seen enjoying the warmer climates of southern Europe, north Africa and even the Caribbean, but does occasionally venture to the UK.

SOURCE: http://tinyurl.com/jxguf38

- **Possible Boob in Navigation by Brown Booby**

A young brown booby (*Sula leucogaster*) was photographed sitting on a trawler just a couple of miles off the coast of Kerry on 13th and 14th August. If genuine, this becomes an Irish (and 'British and Irish') 'first' although it is not a wholly unexpected vagrant, with records from Morocco, the Azores and even southern Spain. Nor is it an unlikely scenario,

boobies being habitual sitters on boats, buoys, pontoons etc. However, doubtless related to higher sea temperatures, warm water seabirds are on the rise around Britain and Ireland, comprising a high proportion of our recent mega-rarities.

SOURCE: http://tinyurl.com/j5tprc6

- **Little Bittern in Bedfordshire**

Here in Britain – Bedfordshire – a juvenile little bittern (*Ixobrychus minutus*) was seen at Marston Vale Millennium Country Park, Bedfordshire on 20th August. This is a major rarity for Bedfordshire, and is only the second county record, the previous one being as long ago as September 1894!

SOURCE: http://tinyurl.com/j5tprc6

- **World's largest protected marine area to shelter millions of penguins**

Antarctica is home to some of the most pristine ecosystems on the planet. It also has fish-rich waters that attract some 46 species of birds. However, even here there is human influence, and penguins are one of the most threatened species due to outside factors such as climate change and overfishing.

The establishment of the Ross Sea Marine Protected Area, which will protect some 1.5 M square kilometres (600,000 square miles) of the Southern Ocean from commercial fishing over the next 35 years has been welcomed by BirdLife.

It is estimated 155,000 Emperor penguin (*Aptenodytes forsteri*) (assessed by BirdLife for the IUCN Red List as Near Threatened) and more than 2.5 million of the near threatened Adelie penguin (*Pygoscelis adeliae*) use these waters. The area is also globally important for the long distance migrant South Polar Skua (*Catharacta maccormicki*) and Southern Fulmar (*Fulmarus glacialoides*) as well as many other species of seabird, and leopard seals, killer whales, nearly a hundred species of fish and approximately 1,000 invertebrate species.

SOURCE: http://tinyurl.com/zakcncs

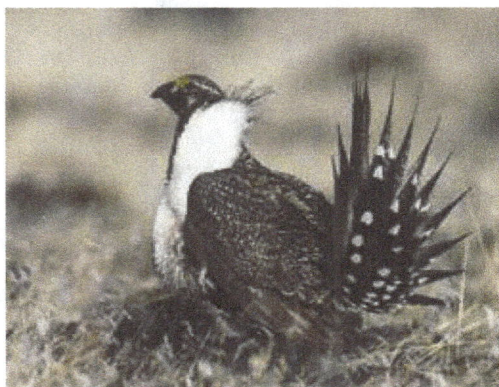

- **Breeding facility for rare endangered bird opens in Calgary Zoo**

Greater sage-grouse (*Centrocercus urophasianus*) were once common in Canada's prairie region, but there are now fewer than 400. Calgary Zoo have opened a state-of-the-art breeding facility to attempt to boost the population of this endangered bird. It is the first of its kind in the country, and there are currently 18 of the birds in this new facility.

Habitat destruction and human development have caused the wild population numbers to greatly diminish. "I see the greater sage-grouse as an iconic part of our Canadian heritage; a key component of our prairie ecosystem," states Dr. Axel Moehrenschlager, director of conservation & science, in a news release provided by The Calgary Zoo.

SOURCE: http://tinyurl.com/hv3nnce

CONSERVATION NEWS

- **Puffin and shearwater numbers surge on Lundy after rat eradication**

More than 300 individual puffins (*Fratercula arctica*) have been counted on Lundy Island this year, compared to just five birds ten years ago, thanks to the Lundy Seabird Recovery Project. Other seabirds have also thrived, including Manx shearwater (*Puffinus puffinus*), with the most recent figures recording some 3,400 breeding pairs, from a low of only 300 pairs when the recovery project was planned and conducted.

SOURCE: http://tinyurl.com/z2c2ohg

- **Reintroducing the Black-fronted Piping-guan in Brazil**

In 2010 it was revealed that only one Black-fronted Piping-guan (*Pipile jacutinga*) was left in the mountain range of Sierra do Mar, São Paulo. SAVE Brasil built a huge enclosure camouflaged in the Atlantic Forest to start a reintroduction programme, and five years later, the bird's plight is being reverted: they are adapting and the locals are making sure their homes stay intact.

This is a globally threatened species endemic to the Atlantic Forest, but as a consequence of poaching and habitat loss, this species is now locally extinct in a big part of its original distribution, such as the Brazilian states of Rio de Janeiro, Espírito Santo and Bahia.

The project team is very excited with release results so far, as the guans seem to be adapting and interacting well with the natural environment. Besides the individual who was sighted far from the enclosure, the others are also being seen in the forest feeding on fruits.

SOURCE: http://tinyurl.com/zcdhuyw

- **Rare birds thriving on Scilly Isles after scheme rids islands of rats**

A scheme to kill rats on two of the Isles of Scilly has led to a resurgence in rare sea birds - the number of Manx shearwaters (*Puffinus puffinus*), has risen to 73 nesting pairs this year, the highest in living memory and almost triple the number of nesting birds just three years ago. Another species of rare ground-nesting birds, storm petrels, have also returned to the Scillies.

There is archeological evidence of Manx shearwaters on the islands dating back to 2,000 BC. By the 13th century, they were so common that they were used as currency. Annual rents were paid in 30 'pufons' (either puffins or Manx shearwaters) to the Duchy of Cornwall.

But it was the rats rather than the Duchy that caused the birds' decline. It is thought that brown rats arrived on the islands in the 17th century, from the many shipwrecks that dot the coast of the Scillies, and by 2014 there were only 24 nesting pairs of Manx shearwaters left and a chick had not survived in some 100 years.

In 2013 the 84 islanders worked together to eradicate the rats under a £750,000 scheme; farmers cleaned out sheds and barns, new, sturdy refuse bins were supplied to every household, and islanders started taking waste to the local tip just once a week.

Plans are underway to extend the scheme to the islands of Tresco, St Martin's and Bryher, if funding can be found. "We know it's feasible," said Jaclyn. "The birds are coming here, so we have a responsibility for them."

SOURCE: http://tinyurl.com/jb66746

- **New Zealand conservationists work to save Rarotonga Flycatcher**

In August, the conservation of Rarotonga flycatcher (*Pomarea dimidiata*) – locally known as Kakerori - and also known as the Rarontonga monarch – got a boost through hands-on training of local staff from visiting predator-control specialists from New Zealand. The bird was formerly common around Rarotonga, yet by the 1900s it was assumed to be extinct. However, in the 1970s and 1980s, surveys found that kakerori persisted in small numbers on the southern side of Rarotonga.

In the spring of 1987, Rod Hay and Hugh Robertson from New Zealand and Cook Islands biodiversity expert Gerald McCormack launched the Kakerori Recovery Programme, under the auspices of the Cook Islands Conservation Service, with volunteers. New Zealander Ed Saul became the backbone of the programme during the third season in 1989, poisoning rats, protecting nests and documenting the efforts. As a result of his continued presence, initially as a volunteer and later as a member of the Cook Islands Conservation Service, the number of Kakerori rose from a low of 29 in his first year to more than 132 at the start of the 1996 breeding season. Today it is estimated there are over 400 Kakerori on Rarotonga.

There is also a population of over 100 on Atiu, where a group of 30 birds had been translocated between 2001 and 2003, as Atiu is free of rats. This was done in order to further protect the species, in case something ever happened to the population on Rarotonga. Atiuan bird expert 'Birdman George' has been instrumental in the protection of these new inhabitants since their arrival on the island.

SOURCE: http://tinyurl.com/gtndxoz

- **Rare cranes breed in Wales for the first time in 400 years**

A pair of cranes bred successfully in Wales for the first time in around 400 years; the pair nested on the Gwent Levels this year, and successfully reared a single chick which flew for the first time in August. The adult birds originate from the Great Crane Project reintroduction scheme which released 93 hand-reared cranes between 2010 and 2014 on the RSPB West Sedgemoor Reserve in Somerset. Damon Bridge, RSPB manager of the Great Crane Project, said: "These wonderful birds died out across the UK sometime in the 1600s, having been a favourite of the medieval dinner table". The Great Crane Project was funded by Viridor Credits Environmental Company and drew on the expertise of the RSPB, the Wildfowl and Wetlands Trust and the Pensthorpe Conservation Trust.

SOURCE: http://www.walesonline.co.uk/business/farming/rare-cranes-breed-wales-first-11981696

- **Crane found shot dead**

The RSPB reported that the body of a female crane released as part of the Great Crane Project five years ago has been found shot dead.

"The bird, nicknamed Swampy, was found dead in a maize field, near Ilchester, by a Somerset farmer recently. The crane's body was handed to the Great Crane Project team and a subsequent post-mortem examination identified four round metal objects identified as gun shot. The most likely cause of death was shooting, the post-mortem examination concluded. Anybody with information should contact Avon and Somerset Police quoting crime reference number 5216228321"

- **Four rare Javan green magpies hatch at Chester Zoo**

Four of the world's rarest magpie hatched at a UK zoo for the first time. The Javan green magpies (*Cissa thalassina*) hatched at Chester Zoo and are listed as a critically endangered species, perhaps even under threat of extinction in the wild, with very few sightings in their southeast Asian habitat. This is mainly due to forest being destroyed for agricultural land, and the birds being targeted by the illegal bird trade in Indonesia.

SOURCE: http://tinyurl.com/h6zbxy6

Other News

- **Flesh-eating mice threaten to wipe out rare birds on wildlife paradise island**

Flesh-eating mice are threatening to wipe out some of the world's rarest birds on a precious wildlife paradise run by Britain. Plagues of hungry rodents are killing up to 600,000 birds a year on one of the UK's 30 UNESCO World Heritage Sites.

Conservationists warned that Gough Island, part of the Tristan da Cunha group, in the South Atlantic is in danger of losing its prestigious status – a standing that puts it on par with the Great Barrier Reef, Yellowstone National Park and the Galapagos archipelago – unless the carnivorous mice are not wiped out.

SOURCE: http://tinyurl.com/zu69cxg

- **Satellite tagged Turtle Doves will reveal migration secrets**

Six turtle doves (*Streptopelia turtur*) are being satellite tracked from their breeding grounds in the UK to their wintering grounds in West Africa, to help scientists better understand why numbers are crashing so rapidly. The number of turtle doves has declined by 93% since 1994, according to the recent UK Breeding Bird Survey.

Last year, in a UK science first, the RSPB revealed the complete migration route of a satellite tagged, UK breeding turtle dove, named Titan, which provided valuable data in the conservation fight to help save the species from UK extinction. Unfortunately, Titan's satellite signal was lost earlier this year when the bird was in Mali and now the RSPB, in partnership with Operation Turtle Dove (OTD), are following six more.

SOURCE: http://tinyurl.com/gup38pg

- **Beijing Cuckoo "Skybomb Bolt" Flies 3,700km Non-stop Across Arabian Sea To Africa**

A common cuckoo (*Cuculus canorus*) fitted with a satellite tag in Beijing in May 2016 made landfall in Africa on the evening of 30th October - a journey of more than 9,000km, the most recent leg of which was a non-stop flight of over 3,700km from central India, across the Arabian Sea, to Somalia.

SOURCE: http://tinyurl.com/j3vwej5

- **The largest hunting violation ever recorded in Jordan**

The illegal killing and trading of birds is one of the rising issues for conservationists in the Mediterranean region. Authorities in Jordan announced recently the seizure of 7,000 dead birds in the largest illegal hunting violation ever recorded in the Kingdom of Jordan after receiving reports about a person who was in possession of large numbers of dead birds.

The rangers from the Royal Department for Protecting Environment and the Royal Society for the Conservation of Nature (RSCN – BirdLife in Jordan) caught the hunter and seized the dead birds in October.

The hunter was found in possession of 6,800 blackcaps (*Sylvia atricapilla*) 40 Eurasian golden orioles (*Oriolus oriolus*) and 45 laughing doves (*Spilopelia senegalensis*).

SOURCE: http://tinyurl.com/zho4gvd

Ten months in the air without landing

Common swifts are known for their impressive aerial abilities, capturing food and nest material while in flight. Now, by attaching data loggers to the birds, researchers reporting in the Cell Press journal *Current Biology* on October 27 have confirmed what some had suspected: common swifts can go for most of the year (10 months!) without ever coming down.

SOURCE: http://tinyurl.com/j6vzqmy

- **Scottish Corncrake numbers fall for second year running**

Populations of one of Scotland's rarest breeding birds, the corncrake (*Crex crex*), have suffered a fall in numbers for the second year running. In total, 1059 calling males were counted during RSPB Scotland's annual survey. That's a drop of 3% when compared with 2015, and a decrease of 20% compared with 2014.

Corncrakes are only found in a few isolated parts of Scotland, mainly on the islands and the far North West coast. This year, the Isle of Tiree was the biggest stronghold, with 346 calling males recorded.

SOURCE: http://tinyurl.com/zwk9wko

- **Azure-winged magpies show human-like generosity**

Magpies do not always have the best reputation, as they are generally known for their tendency to steal shiny things. Also other bird species tested for prosociality so far turned out to be either indifferent to benefitting others or only provided food when the other repeatedly begged for it. Azure-winged magpies (*Cyanopica cyanus*) seem to be the exception to the rule. They provided food to their group members spontaneously and without the other birds begging them. "This so-called 'proactive prosociality' has long been believed to be a human hallmark," explains lead author Lisa Horn. It was however suggested that cooperative child rearing promoted this tendency to benefit others without expecting anything in return in early humans. In line with the hypothesis, researchers also found evidence for prosocial behaviour in cooperatively breeding primates. "But so far results from other animal taxa were missing," says Horn.

Therefore, Horn and colleagues tested prosociality in a cooperatively breeding bird species -- the azure-winged magpie -- using an ingenious experimental design. By landing on a perch, the birds operated a seesaw mechanism, which brought food into reach of their group members. If the birds wanted to grab the food themselves, they would have had to leave the perch and the seesaw would tilt back, thereby moving the food out of reach again. Even though the birds thus could not get anything for themselves the magpies continued to deliver food to their conspecifics across all sessions and at similarly high rates as humans and cooperatively breeding primates. Also, the birds only operated the apparatus when their group members could actually obtain the food, and not in a control condition where access to the food was blocked.

"Our results seem to support the hypothesis that raising offspring cooperatively may have promoted the emergence of prosocial tendencies not only in humans, but also in other animals. Further tests of non-cooperatively breeding birds are, however, needed," concludes Horn.

SOURCE: http://tinyurl.com/z2q6mhx

- **Seabirds eat plastic because it smells like food**

Very little research examines why birds make the mistake of eating plastic.

Now a study, in the journal *Science*

Advances, helps explain why plastic ingestion is more prevalent in some seabird species than in others. Tubenosed seabirds, such as petrels and albatross, have a keen sense of smell, which they use to hunt. They are also among the birds most severely affected by plastic consumption. Marine plastic debris emits the scent of a sulfurous compound that some seabirds have relied upon for thousands of years to tell them where to find food, according to a study from the University of California, Davis. This olfactory cue essentially tricks the birds into confusing marine plastic with food.

"It's important to consider the organism's point of view in questions like this," said lead author Matthew Savoca, who performed the study as a graduate student in the lab of UC Davis professor Gabrielle Nevitt and who is with the Graduate Group in Ecology.

"Animals usually have a reason for the decisions they make. If we want to truly understand why animals are eating plastic in the ocean, we have to think about how animals find food."

SOURCE: http://tinyurl.com/jgbubem

- **Cooperation between honeyguide and humans is a two-way conversation**

People in Africa are able to locate bees' nests to harvest honey, by following greater honeyguides (*Indicator indicator*) and by using special calls to solicit their help. Research also suggest that greater honeyguides actively recruit appropriate human partners. This relationship is a rare example of cooperation between humans and free-living animals.

Honeyguides give a special call to attract

people's attention, then fly from tree to tree to indicate the direction of a bees' nest. We humans are useful collaborators to honeyguides because of our ability to subdue stinging bees with smoke and chop open their nest, providing wax for the honeyguide and honey for ourselves.

.

SOURCE: http://tinyurl.com/gpmxmbr

- **Songbirds' epic migrations connected to a small cluster of genes**

Scientists from the University of British Columbia have shown that there is a genetic basis to the migratory routes flown by songbirds, and have narrowed in on a relatively small cluster of genes that may govern the behaviour.

"It's amazing that the routes and timing of such complex behaviour could be genetically determined and associated with a very small portion of the genome," said researcher Kira Delmore, lead author of the paper published in *Current Biology*. "What's even more amazing is that differences in this behaviour could be helping to maintain the huge diversity of songbirds we see in the natural world."

SOURCE: http://tinyurl.com/hsh22dc

- **UK first as Brecks Stone Curlews are GPS tagged**

For the first time in the UK, scientists working in the Brecks are using high-tech GPS tags to study the movements of one of the country's most threatened birds, the stone curlew.

Stone curlews *Burhinus oedicnemus*, were

close to becoming extinct as breeding birds in the UK 30 years ago, but thanks to conservation efforts, around 400 pairs now breed in the UK each year – more than half of those in Eastern England. By using GPS tracking to learn more about how these shy and elusive birds use different areas of the countryside, researchers hope to help landowners create the conditions Stone Curlews need for nesting and feeding, in order to ultimately achieve a sustainable Stone Curlew population in the UK.

SOURCE: http://tinyurl.com/j8mh24j

- **Scientists find first evidence of birds sleeping during flight**

For the first time, researchers have discovered that birds can sleep in flight. Together with an international team of colleagues, Niels Rattenborg from the Max Planck Institute for Ornithology in Seewiesen measured the brain activity of frigatebirds and found that they sleep in flight with either one cerebral hemisphere at a time or both hemispheres simultaneously. Despite being able to engage in all types of sleep in flight, the birds slept less than an hour a day, a mere fraction of the time spent sleeping on land. How frigatebirds are able to perform adaptively on such little sleep remains a mystery. Researchers found that, on average, great frigatebirds (*Fregata minor*) sleep for 42 minutes per day when on the wing.

SOURCE: http://tinyurl.com/z2grjx9

- **India will lose all Great Indian Bustards by 2020**

Conservationist Bikram Grewal's book includes some 1,300 species of birds found in India and its neighbouring nations, many of which are facing extinction in his country; he fears that all bird species are endangered. "We are losing everywhere, population of common urban birds, migratory birds and rare birds have only witnessed a decline in India," says the eminent birdwatcher. Without beating around the bush, he blames complete lack of political will. "We don't need high-level science to save birds. Just had to keep their habitat sacrosanct but we couldn't."

This is exactly what happened with the Great Indian Bustard (*Ardeotis nigriceps*), he says. "There are hardly 50 of them left in the entire country. 90% of the species are in Rajasthan, where its habitat is being drastically destroyed. By 2020, GIBs will be extinct," adds Grewal. Though the situation is critical now, he adds that a lot can still be done to save the remaining population of the bird. "When villages can be relocated to protect tigers, similar efforts can also be taken for prominent bird habitats," he says.

SOURCE: http://tinyurl.com/zqny5lt

- **Rare bird being driven to extinction by poaching for its 'red ivory' bill**

The solid red beak of the helmeted hornbill (*Rhinoplax vigil*), known as a casque, sell for several times the price of elephant ivory due to soaring demand on the Chinese black market, and this poaching is driving the bird to extinction.

The helmeted hornbill is found mainly in Indonesia, Borneo and Thailand, and the huge birds have been caught for centuries for their tail feathers, prized by local communities. However, since 2011 poaching has soared to feed Chinese demand for carving ivory, even though the trade is illegal, sending the hornbill into a death spiral.

SOURCE: http://tinyurl.com/hl3a4rp

- **Hunted to the brink, but Africa's reviled vultures are vital in fight against disease**

Vultures are rarely viewed as the poster boys and girls of the natural world. They have repulsive eating habits and are strikingly ugly. Nevertheless, they play a critical role in maintaining the ecological health of many parts of the world by consuming animal carcasses more effectively than any other scavengers, and because their digestive juices contain acids that neutralise pathogens such as cholera and rabies they prevent diseases spreading. Vultures are one of the fastest declining groups of animals in the world. In India, all nine species of the bird are threatened with extinction, largely through the indiscriminate use of diclofenac, a common anti-inflammatory drug administered to livestock but which is lethal for the vultures that eat the corpses of cattle.

"There is now a real danger that a disease like rabies will spread because there are hardly any vultures left to clean up corpses left in the open," photographer Hamilton James said.

SOURCE: http://tinyurl.com/h8njju7

- **Natural England issues licence to kill Buzzards to protect pheasants**

In July Natural England announced that it had issued a licence for someone to kill up to ten Buzzards "to prevent serious damage to young pheasants".

Here is their statement:

29 July 2016
Natural England issued a licence last night permitting the control of up to 10 Buzzards to prevent serious damage to young pheasants.

The licence is time-limited with stringent conditions and is based on the law, policy and best available evidence. It follows rigorous assessment after other methods had been tried unsuccessfully over a 5-year period.

It is stipulated that the licence must be used in combination with non-lethal measures and only on Buzzards in and immediately around the animal pens – not on passing birds. These conditions are designed to make the licensed activity both proportionate and effective and we will continue to work with the applicant to assess this.

Killing wild birds without a licence from Natural England is illegal.

SOURCES: http://tinyurl.com/jfsdtha
                    http://tinyurl.com/jb4hgwo

- **Hornsea offshore wind farm decision devastating for iconic seabirds**

The Secretary of State's decision to approve the Hornsea Project Two offshore wind farm has concerned the RSPB in that it will lead to the unnecessary death of hundreds of globally important seabirds. Although the RSPB supports the move to generating more electricity through renewable sources and encourages the development of renewable energy projects, the charity feels that this must be delivered in harmony with nature. This means carefully looking at each site and the potential impact or risks any proposal may have on local or migrating wildlife. Unfortunately, the Government licensed this area for wind farm development without doing the necessary surveys.

SOURCE: http://tinyurl.com/zcpy7gc

- **Birds fly faster in large flocks**

New research at Lund University in Sweden shows that the flight speed of birds is determined by a variety of factors. Among the most sensational is that the size of the flock has a significant impact on how fast the birds can fly. The larger the flock, the higher the speed.

Researchers at the Faculty of Science in Lund have now shown how several factors, working simultaneously together, determine the birds' flight speed. Their morphology, that is, the bird's weight and the shape of its wings, is one factor; wind direction and speed is another; and the situation (searching for food or travelling long distances) is a third.

However, what surprised the researchers the most was that the flock size has a major impact on the birds' speed.

SOURCE: http://tinyurl.com/hbo3v8t

- **Welsh Government refuses to ban shooting of White-fronted Geese**

Greenland white-fronted geese (*Anser albifrons flavirostris*) are endangered and under serious threat of extinction with the global count now being below 20,000 birds. The wintering population in Wales is at critically low levels; in the late 1990s, over 160 birds returned to their regular wintering site on the Dyfi estuary, but that figure was down to only 24 last year with a small number of sightings at one or two other places in north Wales. However, the Welsh Government has announced that it will not implement a total ban on shooting white-fronted geese across Wales.

According to RSPB Cymru Director, Katie-jo Luxton, "When a species is declining so quickly that it is under threat of extinction, you'd think the least that those in power could do is to offer it legal protection to prevent it from being shot.

It is not clear how the Cabinet Secretary for Environment and Rural Affairs came to this decision, but what we do know is that she has gone against the advice of Natural Resources Wales, Welsh Government's own statutory nature conservation and environmental regulator, as well as the opinions of a number of Wales' conservation organisations and a large numbers of individuals."

SOURCE http://tinyurl.com/gnmqpwb

For those of you not aware, as well as this column in *Animals & Men,* Corinna writes a daily Fortean bird blog which can be found as part of the CFZ Blog Network, but also as a stand alone site at:

http://cfzwatcheroftheskies.blogspot.com/

PULL OF THE BUSH: BACK ON THE TRACK OF THE TASMANIAN WOLF

In February of 2016 I returned to Tasmania for my second attempt to find the Tasmanian wolf, or thylacine, the iconic marsupial predator that is the emblem of the Centre for Fortean Zoology. The last expedition I had taken part in, back in 2013, had consisted of many people from Australia and the UK. On this occasion, Mike Williams of CFZ Australia had decided, wisely, to pare it down. This time it would be a skeleton crew of Mike and myself on the track of the legendary beast.

Flying to Tasmania is a long affair, taking a day or more, and consisting of three changes. Getting to your destination is the only leg of an expedition that really worries me. Once I'm in the field I'm fine. I finally got to Launceston and was met by Mike. Over a coffee he explained some developments since I'd last been to Tasmania.

I'd missed the 2015 expedition through a bout of gout and pneumonia. On that trip the team had been in the north east of the island, but met with less success than on the 2013 trip in the north west. The area we had visited on that trip had been subject to

RICHARD FREEMAN

savage bush fires; multiple lightning strikes had caused fires to reduce much of the forests to ash. It's a natural process in Australia, but it rendered the area less than perfect for our purposes.

Mike had been in contact with a farmer in the north east, who said he had captured a thylacine on camera. The man, who wanted to remain nameless, had allowed Mike to look at the pictures after much persuasion, but would not allow copies to be made. Mike was convinced of the authenticity of the pictures, mainly due to one interesting feature; the creature in the two pictures had a shaggy winter coat. Most people do not realize that the Tasmanian wolf grew longer hair in the winter months, and most reconstructions of them show the animal with a short coat.

The first picture showed the creature side on to the camera trap, and the second showed it turning away. The stiff tail and stripes were apparent. The farmer had placed the camera trap on a hill on his property for months on end. These were the only two pictures he had gotten over that period. He was cagey about showing them to anyone else, or to being interviewed.

Mike had heard of some recent sightings further south on the island, and with our former area being burned out we decided to make this our HQ for the trip. As before, we decided not to reveal the exact location of the sightings in order to protect the animals.

We camped out at a grassy area with wooded hills on the first night. We found a dead Tasmanian devil on the road, which had apparently been hit by a car. There were no signs of the facial tumours that have been ravaging the population elsewhere.

We had brought camera traps with us, and we affixed these to trees in remote areas, and used road kill as bait. In addition to this, we once again employed bonnet mounted cameras that film constantly as we did our night drives. Most sightings of the Tasmanian wolf are made by motorists at night. Should anything run in front of our vehicle it would be caught on film.

Once more Toyota had generously lent us a four-wheel drive car for our trip.

Next day we crossed the Western Tiers, a beautiful escarpment studded with lakes. We made camp and set up cameras. By day we explored on foot and by night we drove the remote country roads. Bennett's wallaby, pademelon, wombat and eastern quoll were all in abundance. The tiger quoll and the Tasmanian tiger were not nearly as apparent as they had been further north.

The following day we drove to a small town to meet our first witness, Joe Booth. When Mike and I arrived, Joe was in his garage trying out a home-made prosthetic hand that appeared to have been made from a sharpened curtain hook and an old aerosol can. Greeting us enthusiastically, Joe, who was an instantly likeable bloke, explained about his home-made hook. The year before he had been out with his mate who was a keen hunter. Joe had been standing outside his mate's car. On the back seat was a dangerous combination of loaded guns and excitable dogs. As the dogs bounded about one of them knocked the guns which had the safety catches off. One went off blowing a hole through the side of the car. It also took a chunk out of Joe's side, and blew off his right hand.

Joe was lucky to survive and had to have several transfusions. However, he bore his mate no grudge and seemed to take his disability in his stride and did not let it affect him in the least. He found the prosthetic hand given to him by the hospital uncomfortable and got on better with the one knocked up at home.

Joe had been a logger in the '50s, '60s and '70s and had seen some of the most massive trees in Tasmania fall to the power saw. He told us that in remote areas the crew would regularly come upon dog-like tracks. He asked the foreman who on the crew had a dog? The foreman replied that they were the tracks of a Tasmanian tiger. One of the other workers scoffed at the idea. A few days later the same man walked around a large tree stump, and found a thylacine sitting there. The animal gave a warning gape and the man backed swiftly away.

In the 1950s, Joe had his own sighting. One evening he was putting his car away. It was twilight and he saw what he thought was his neighbour's dog walking down the road towards him. He called out to it but it didn't react. As it drew closer and walked past him he saw it had a thick, stiff tail and stripes along its hind quarters. He then realized that he had seen a thylacine. He had recalled hearing that a crop sprayer pilot had said that he had seen one in the vicinity some days

earlier. A few days after Joe's sighting, one of his mates who lived locally saw the creature. It ran out of a woodpile and vanished between some barns. This is interesting, as thylacines were thought to make temporary dens that they used for a few days before moving on

Joe's wife Pat had also seen the Tasmanian wolf. Thirty-five years ago, in 1981, she had been driving a couple of miles outside of town. It was winter and twilight. A Tasmanian wolf crossed the road in front of her car. She got to within fifteen feet of it. She clearly saw the striped flank and stiff tail. It was 18 inches to 2 feet tall with a yellowish brown coat and powerful looking jaws. It was somewhat greyhound-like. Pat had it in view for 60 seconds before it moved off into the surrounding fields.

One of Joe's interests are the old convict roads. These were constructed by convicts transported to Tasmania from the 1830s onwards. He and a number of friends try to locate and restore the roads. He took us out to show us a rock that had a bizarre carving on it. It had been made by one Nehemiah Rogers. Originally from Brocking in Essex, and a stonemason by trade, Rogers was born in 1825. Convicted of burglary in 1845 he was transported to Tasmania. Joe didn't know what the strange symbol carved into the boulder represented, and thought it might have been masonic. To me it looked like a stylized, ejaculating phallus.

Joe explained that the previous year he and his son had been exploring the wooded hills some miles from the town. His son had been on a gravel path and Joe had been deep in the undergrowth some way from him. Stumbling across some ruined huts Joe had called out to his son. Apparently his shouting had disturbed an animal. His son shouted out to him that a strange animal had emerged from the bush and was on the road a few yards ahead of him. By the time Joe had got to the path the animal had gone. His son described it as the size of a whippet with tan coloured hair, dark stripes on the sides and a stiff tail. It trotted off up the track. The creature, apparently a young thylacine, had left a set of clear tracks. Joe and his son followed them up the road till they vanished back into the bush. On returning to their car, it seemed that the creature had doubled back and walked around the vehicle before returning to the forest.

Joe returned next day with a camera and photographed the paw prints. They seem to show five visible claw marks on the front foot. The Tasmanian wolf was plantigrade, unlike the placental wolf that was digigrade.

This means that it walked on the whole of the foot and not up on the toes like true dogs. The dog's dew claw equates to our thumb or big toe, and is held clear of the ground. Clear tracks of a thylacine's front foot generally shows five claw marks, a dog will show four. Also there was a small indentation behind the metacarpal pad (that equates to the palm) on each print. Again this is typical of a thylacine.

Mike and I made camp in the area, and set up camera traps bated with fresh road kill, or oven ready chickens. We spent the day exploring on foot, and the nights driving.

The following day we travelled to another town in the area to meet up with veteran thylacine hunter Col Bailey. Col saw a thylacine back in the 1960s on the mainland,

whilst on a canoe trip in the Coorong Lakes in 1967.

"400 yards away I saw a dog-like animal on the water's edge. It was big, like a greyhound, a long animal with short legs, a long tail and a big head. But then it disappeared."

This fired his interest in the animal and he moved to Tasmania. Col was lucky enough to meet and interview old bushmen who had been around in the late 19th and early 20th centuries, and mine their wealth of knowledge on the Tasmanian wolf. Without Col's work and diligence these stories and information would be lost to the ages as all the old trappers and bushmen have long since passed away.

Col saw the animal again in 1995, this time on Tasmania and in deep bush.

"My eyes ran down its back and tail and it hit me — this was clearly a Tasmanian tiger. I was entranced, riveted to the spot. I stood there and watched it for almost a minute before it hissed at me and turned into the bush."

Beforehand he had heard the distinctive high pitched yap of the animal and smelled its pungent odour.

Col kept the sighting under his hat for 17 years in order to protect the creature.

Col, now 78, has written three books on the thylacine, *Tiger Tales*, *Shadow of the Thylacine* and most recently *Lure of the Thylacine*.

We spoke for some time, and covered the power of the animal's bite. A recent paper

tried to claim that the animal had weak jaws and would only feed on small creatures like possums. This is totally at odds with field reports at the time, which said the Tasmanian wolf killed and ate kangaroos, wallabies and full grown sheep, killing them with exceptionally powerful bites. Several reports said that when cornered by dogs, a thylacine could bite clean through a dog's skull. A more recent paper refuted the weak jaw hypothesis. Looking at the skull anatomy, it's authors concluded that the thylacine had a much more powerful bite than a wolf or dog, but the skull was not as well adapted to hold on to struggling prey. Wolves, being pack hunters, surround their quarry and hang on to it, worrying it to death. The solitary thylacine kills with one or more powerful bites.

Col also spoke of the absurd numbers of sheep kills laid at the thylacine's door in the bounty years. It would have been impossible for the animals to have killed that many sheep without attacking them 24/7.

Next day we visited a remote valley area before returning to Joe's town. The librarian there had a story to tell. Twenty years before, her car had broken down some miles outside of town. There were no lights and she was compelled to follow the road in darkness towards the town. She soon became aware of a soft padding behind her. It was too dark to see anything, but she knew that something was following her. She shouted out and whatever it was ran back into the bush. Later she read of how Tasmanian wolves would often follow men in the bush. She was convinced that it was one such creature that tracked her that night.

We drove down to Hobart to see the thylacine display at the museum. There were stuffed specimens, pelts, skulls, bones and casts of the last known prints taken in the wild (according to them) but not a word on thylacine survival, or the 4,000 plus sightings since the 1930s.

Later we checked the camera traps. One of the bait carcasses had been partially eaten. A hole was ripped behind the back leg and the internal organs had been devoured. Something had taken the head too. Looking at the pictures we saw two quolls and a Tasmania devil.

The following day we returned to Joe's to borrow the photographs and make copies. Joe told us of some men out spotlighting who had seen a thylacine just six months before.

He also told us about the shack that was once home to Elias Churchill, a trapper who captured thylacines alive for zoos in the early 20th century. Col Bailey rediscovered the shanty, not used since the early 1930s, back in 2006. The hut was restored with a grant from Tourism Tasmania. Mike and I decided to take a look at it.

The location was remote and quite a distance away. Following directions and hand drawn maps we found ourselves along a track in a wooded area, but we failed to locate the hut. Mike walked on ahead and I lingered behind him. On a section of the track I became aware of a weird smell, somewhat like that of a hyena. I am a former zookeeper and regular zoo visitor and I am familiar with the smell. The odour seemed to intersect the track and was only in one area. It was as if whatever had left the scent behind had recently crossed the track. The Tasmanian wolf was said to smell very much like a hyena.

We tried to find the shack again the next

day, and this time managed to get to it. The hut was small and I was impressed that Churchill weathered the harsh Tasmanian winters in the structure. The remains of the stockade where he kept captured thylacines was still standing as well. Churchill snared them, kept them in the pen then transported them out of the wilderness on horseback.

We placed a camera trap at the area where the odd smell was detected. On the way back I found some scat and preserved it in ethanol for analysis. It was dark and fudgy, matching the description of thylacine droppings as the content of their diet is rich in blood. The dropping that we had found on the last expedition were found to contain bone fragments and were ultimately shown to be from large Tasmanian devils. This sample looked very different, lacking bone chips but containing hair.

Driving down to the south west, we visited Lake Pedder, Australia's largest freshwater lake. It was once a natural lake of modest size. In 1972 The Hydro Electric Commission of Tasmania flooded the lake by damming the Serpentine and Huron rivers and extending the lake to its current size of 242 square miles. The project was opposed by conservationists and galvanized the green movement in Tasmania. Tasmanian premier Eric Reece supported the project and gave the following, appalling quote.

"There was a National Park out there, but I can't remember exactly where it was ... at least, it wasn't of substantial significance in the scheme of things."

In 1972, the activist Brenda Hean and pilot Max Price were killed when their Tiger Moth plane crashed. They were flying from Tasmania to Canberra to protest the damming of Lake Pedder; it was alleged that pro-dam campaigners had entered the plane's hangar and placed sugar in one of its fuel tanks.

The flooding led to the extinction of the Lake Pedder earthworm *(Hypolimnus pedderensis)*. Another victim was the Pedder galaxias *(Galaxias pedderensis)* a tiny fish found only in the lake. It is now extinct in the area, but populations have been translocated to one at Lake Oberon in the Western Arthurs mountain range, and one at a modified water supply dam near Strathgordon.

Sickeningly, big business always seems to triumph over environmental or conservation concerns. Though it looks beautiful, the lake leaves a bad taste in the mouth. There are pressure groups today that are advocating the draining of the man-made lake and the restoration of the original Lake Pedder.

Joe had told us of another local man, Bill Morgan, who had seen a Tasmanian wolf back in the 1970s. We tracked him down and he agreed to talk to us.

A sprightly 93-year-old, Bill had worked for the hydro-electric company. He encountered a thylacine in 1979 but would not reveal the exact location. Bill was in a carful of co-workers. They drove over a bridge and saw a thylacine in the middle of the road. Bill described the animal as 'beautiful' with sleek fur and stripes. It moved with stiff looking hindquarters. It left the road and looked back at them as it went. The group had it in view for six minutes.

His friend, Max Macallum, also saw a thylacine in the same year. The animal crossed the road in front of him as he was driving to his brother's house.

Bill had recently caught up with his cousin whom had had not seen in decades. Amazingly, just eighteen months before, the cousin and five other people in a car, had seen a family group of thylacines. A male, female, and three pups crossed the road in front of them. It happened on the road to a town in West Tasmania. Bill had no doubt that the Tasmanian wolf was still around.

We visited a range of mountains in which there are the remains of an old osmiridium mining town. Osmiridium is a natural alloy of osmium and iridium used mainly to make pen nibs. Tasmania was the world's foremost supplier of the alloy. Only a few preserved shanties remain of the town.

We checked the camera traps and found only devil, quoll and other fairly common creatures

R20   03-08-2016   10:34:23   2/3   46F   O

on the pictures. The traps were re-baited with fresh meat.

We took time out to visit an artist called David Hurst. He is producing life-sized bronze busts of thylacine heads. He showed us his workshop where he was first carving the heads in wax. They were remarkable in detail. David thinks the animal is still with us, and thought that the south west wilderness might prove a bountiful area.

We visited Joe again, and we headed up to the hills once more. He told us of finding the remains of an aboriginal hearth under a felled tree in the early 1970s. He believed that the hearth had been preserved there for over fifty years.

We met with Kathy Brownie, the proprietor of a local coffee shop. She had played a bit part in the 2011 film *Hunter* starring Willem Dafoe, Sam Neill and Frances O'Connor. The film sees Dafoe as a hunter employed by a pharmaceutical company to track down the thylacine. An unimpressive flick, it is filled with scientific errors such as giving the animal a venomous bite! Much more interesting was the small museum she maintained in the shop. Among the fossils and minerals were alleged casts of the hind foot of a Tasmanian wolf. It clearly showed the long carpal pad. The casts of the tracks were taken back in 1991 by a guy called Rusty Morley. Kathy told us that in 1971 she was living in a mining town consisting of wooden shacks and very limited amenities. She said that a bulldozer had uncovered an old thylacine lair that had the

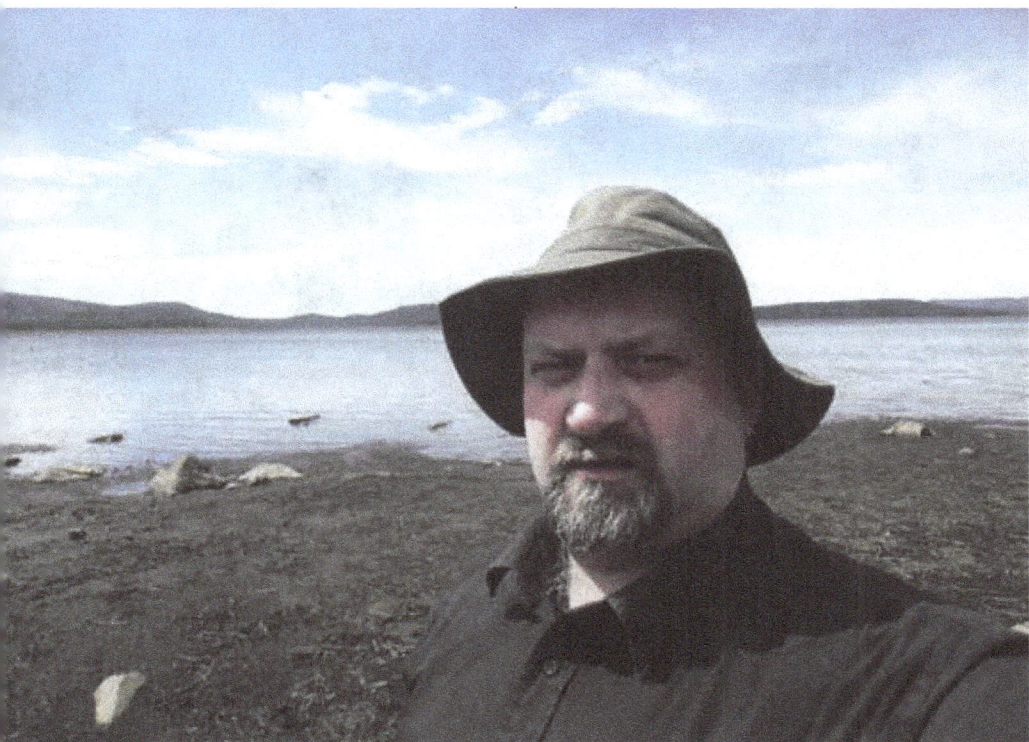

remains of prey animals in it. She claimed to have heard the Tasmanian wolf's call on a number of occasions.

Upon returning to his home, we found that seven pigs, escapees from some neighbouring farm, had chomped their way through Joe's potato plants and were now making free with his pumpkin patch. I chased them out of the garden and down the lane with a mop.

As the expedition wound down, we returned to Launceston and visited the natural history museum there. We found it to be better than the one at Hobart. It covered the possibility of thylacine survival and had a map of sightings from the 1030s to the early 1970s. Why this had not been updated was anyone's guess. The next expedition is provisionally scheduled for February of 2017. We may be returning to the same area or pushing deep into the uninhabited south west of the island, an endeavour that may necessitate travel by boat.

Once again we had spoken with witnesses who had no axe to grind. Many intimately knew the Tasmanian bush. The Tasmanian wolf is still reported each year. There are many iconic extinct animals; the dodo, the great auk, the Bali tiger, the moa, the passenger pigeon, and the Carolina parakeet. Nobody reports seeing any of these animals because they are extinct. People still report seeing the Tasmanian wolf. If it really were extinct then nobody would be seeing it. The thylacine does indeed merit the title 'the healthiest extinct animal you will ever see.'

Plate 1

TROGLODYTES GORILLA. man adult.

THE GORILLA - DOES IT EXIST?

This is the title of a chapter in the book *'Sea and Land'* by **James William Buel** (1849-1920). The book of 800 pages was published in 1887, in times when the knowledge of

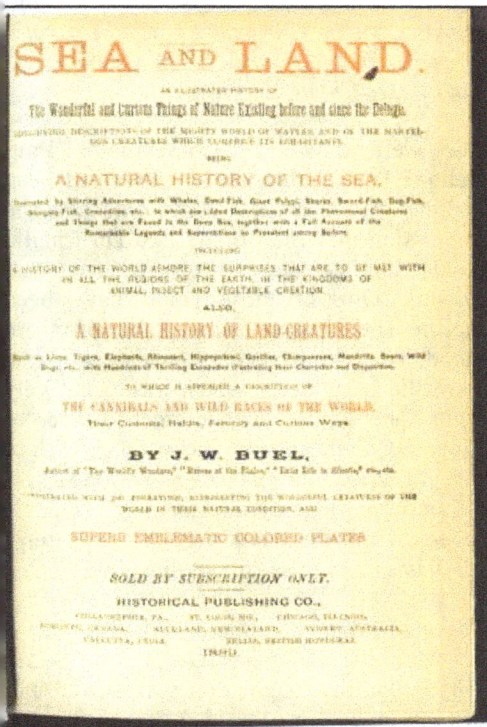

many tropical animals was brought to the West by explorers who were courageous enough to risk their lives in remote places, hacking and shooting, for nature in general and the jungle in particular was the enemy of civilization.

Buel is very pre-occupied with the above question. His book is a systematic treatise on the dangers of nature, the 'ferocious animals' in the past and present, and the human methods of effectively dealing with those dangers, by killing the beasts. About 300 engravings enlighten the reader how that was done.

Chapter XXXIII is about primates. The existence of the orang utan had been confirmed since the 17th century. Living specimens entertained the public in appalling cramped accommodations in the oldest zoos of the 19th and 20th century and their remains entertained scientists and the public alike. The identity of a newly discovered ape, especially when there was no life specimen available, could be a reason for discussion.

So it was with the gorilla.

LOES MODDERMAN

MEETING THE GORILLA.

traveler who so much as pretends to have seen the creature, or even to have heard of it through the wild tribes of that country, is Paul B. Du Chaillu."

I'll cite from Buel's book. Buel himself was far from sure if something like the gorilla really existed, even when by the end of the 19th century most zoologists seemed to accept there was such a creature.

Buel starts mentioning the unicorn, of all animals, *"which, many years ago, was fiercely debated by every learned person in Europe. It has at length been settled that the gorilla is a reality, while the unicorn is a myth, but there are not wanting persons who, if not still openly disputing the claim, do entertain grave doubts as to the existence of the gorilla, and the reasons for so believing are not without some force. It is a most astonishing fact that the only African*

The well known French explorer **Paul Belloni Du Chaillu** (1831-1903) had been making waves in those days. One gets the impression that Buel didn't like Du Chaillu much, but he doesn't tell us his exact reasons. One of those could have been racism. The explorer had a mother of mixed blood, something very inconvenient in the scientific community, which was exclusively white and male. An internet article of Richard Conniff, *Race, Sex and the Trials of a Young Explorer* http://opinionator.blogs.nytimes.com/2011/02/13/race-sex-and-the-trials-a-young-explorer/?_r=0 has more to say about this issue.

The main thing why the gorilla was such a hot item was Darwin's *Origin of Species* (1859). Public imagination was stirred and often infuriated by the idea of humans and apes being related. The discovery of the gorilla, of all primates anatomically nearest to humans, fuelled the scientific debate how animals and men are different, physically and in brain function. Books were written about the question which ape was our nearest cousin. The chimp was also a worthy candidate. Science was divided about it, religion dismissed Darwin with a vengeance. The champion of evolution was ridiculed in a multitude of cartoons.

But the general public was fascinated by the discovery of the gorilla and by the adventures of Paul Du Chaillu.

From 1856 to 1859 Du Chaillu travelled through western Africa, bringing back a choice of plants and animals, 20 gorilla skeletons and skulls among them. In 1861 he was nothing less than a celebrity when he addressed the *Royal Geographical Society* of London, telling about his many adventures with a row of stuffed gorilla's to back him up. But his many enemies were there too. They thought him an amateurist. Obviously exploring 'on the spot' carried much less weight than being a scientist who never set foot outside a scientific institution.

It is unlikely that Buel, only 12 years old in 1861, had personally heard Du Chaillu lecturing, but he certainly grew up with the gorilla, which could hardly be missed in the second half of the 19th century.

The man Buel is kind of a mystery. No picture of him anywhere. Apart from *Sea and Land* he wrote several other richly illustrated books, *The Story of Man* (1889) among them. But what his own profession was I can't find on the internet. He certainly had a love for 'horrible creatures', the 'alleged' gorilla among them. Maybe that's why he also wrote a book on notorious criminals.

Sea and Land was published in 1887, about 25 years after Du Chaillu's heydays as the man who discovered the gorilla. One should think that the gentlemen of science would have made up their minds about the distinction between the gorilla and the orang

utan by then. But we don't know when exactly Buel wrote his evaluation of Du Chaillu's discoveries, and if he ever saw them himself.

He writes: *"Without throwing discredit on Du Chaillu's statements* (which he keeps doing), *it is plain that naturalists have too promptly accepted the evidences which he has produced. The skeletons, it is true, speak for themselves, as it were, but hardly conclusively. The principle point to decide is whether the Orang Outan and the gorilla are not one and the same."*

He then names several other explorers, Livingstone among them, *"who hunted for years through the same country, but never so much as heard of the creature. This fact in itself is singular."*

Buel's position is clear. At first he throws his doubt between the lines but gradually he makes it more openly clear that he doesn't believe a word of Du Chaillu's claims.

He quotes the explorer who describes the gorilla as *'extremely fierce, attacking a man, if pressed, with such extraordinary ferocity that, unless it be immediately killed by a well directed shot, the hunter is sure to fall a victim before it. '*

Buel mentions that Du Chaillu only brought two skeletons of gorillas with him, while other sources mention 20.

They resemble, Buel keeps saying, the orang utan, except the skulls. Then, like a genuine enthusiast for explaining away the obvious, he tells us that there is no proof in that, for people also have different craniums and why shouldn't orang utans? *"An injury, disease, brain development, food, and a hundred other things, not to speak of malformations of birth, may account for the differences*

MY FIRST GORILLA.

STORIES OF THE GORILLA COUNTRY

PAUL DU CHAILLU

which we observe in the skulls of Du Chaillu's two specimens and those of the authenticated Orang Outan. I do not say that the gorilla is a myth, but until stronger evidences of its existence are produced we may expect that there will be doubts of its existence. "

How many times can a man keep expressing his doubt?

Buel reminds us of the scientists who explained *Homo floresienses*, the 'hobbit', when it was discovered in 2003, as an isolated case of malformation, a growth defect; everything better than contemplate the 'impossibility' of a new humanoid species, and a recent one at that.

He then continues with several exaggarated reports from travellers in Africa, one of which told stories of apes gathering the fallen tusks of elephants (oh? since when do they drop spontaneously?) and more such strangeness.

One explorer knew of big apes walking on their hindlegs.

It's not the elephant tusks in the jungle that make Buel suspicious, it's the walking on hindlegs, as that can't be. Only humans can do that. Beasts are created to walk on all fours.

Buel tells us that he has *"patiently gone through no less than eleven hundred (!!) different books on Africa and found nowhere, save in Du Chaillu's works, any description of the gorilla."*

That man was motivated!

Buel cites a lot from a book of Du Chaillu, who describes his courageous hunt for gorillas in colourful language.

Although the man admits that gorillas have human-like traits, that doesn't stir his sympathy.

At a certain moment he hides in the bushes to observe the animals. They don't see him and they can't smell him.

He says: *"How fiendish their looks were! A cold shiver ran through me several times, for, of all the malignant expressions I had ever seen, their's were the most diabolical"*

It's very hard for any of us who love animals to understand how a natural animal can look 'malignant' or 'diabolical'. But such simple expressions convey the deep gap that separates us from how nature was experienced in the 19th century.

Du Chaillu then tells us about a 'present' he received when coming back from an expedition: a female gorilla, shot, bound, with a fractured skull and a broken arm, with on her belly a baby. By her side is a male gorilla, watching over what once was his thriving family.

The explorer leaves the dying animal on his doorstep, does nothing to help her and nothing to help the baby, and he doesn't shoot her, what was maybe the most decent thing to do. The next morning she still lives. He admits that 'her pitiable cries' kept him awake at night. And he wonders about God who made these creatures 'almost' - but not quite - like humans.

The mother dies, and the child dies too, after a while, as does another captive child gorilla.

Du Chaillu takes a picture of the beasts , an 'excellent one' .

Buel then goes back to his own story: how he as a child saw a monstrous stuffed ape in one of Barnum's sideshows, and concluded that that must have been a gorilla *"because it was too ugly to be anything else"*

And after reminding us of the mermaids and unicorns made to deceive, he says: *"I therefore repeat, that until the gorilla is brought under the examination of approved scientists who are familiar at least with the Simian family, cool- headed people, while not denying, will not conclude positively that there is in Africa, or in any other part of the world, such an animal."*

One wonders if there hasn't been ample time for 'cool headed approved scientists' to examine the evidence?

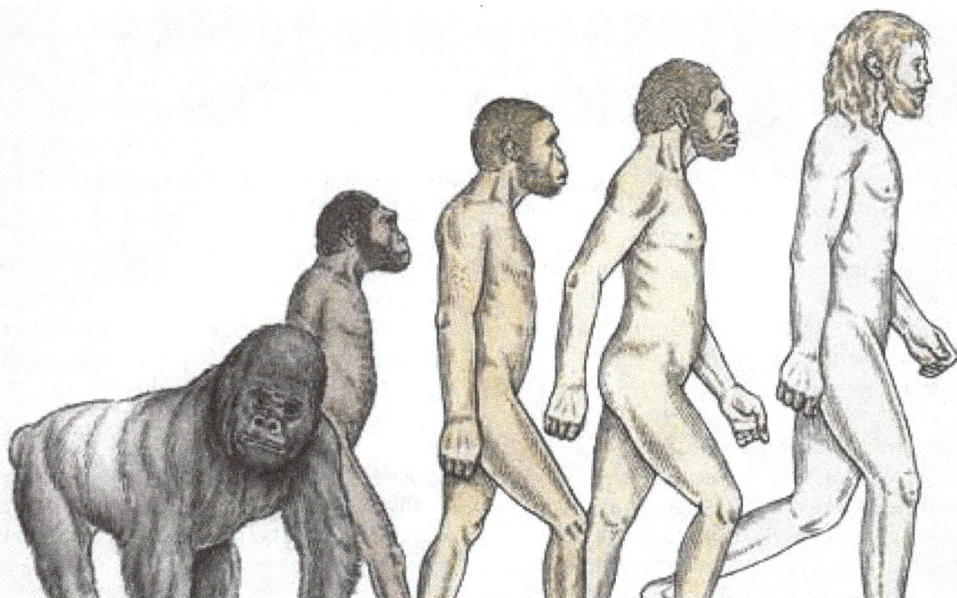

Approved by whom? By Buel?

He ends his longwinded argumentation, before proceeding to the habits of the orang utan, with a last sneer in the direction of Du Chaillu: *"It is much to be regretted, if such an animal really exists, that we have to depend upon a single authority for all the knowledge we possess concerning it, for all the interesting products of creation surely none exceed the gorilla, and the naturalist is constantly*

thirsting for more information respecting it."

I wonder if Buel ever got convinced the gorilla exists?

About Du Chaillu http://www.mainlesson.com/displayauthor.php?author=chaillu

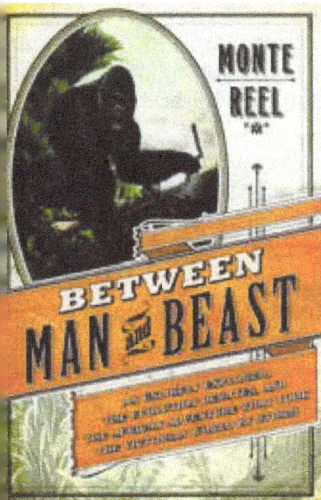

DISCUSSION DOCUMENT: DUTIES FOR REGIONAL REPRESENTATIVES

The CFZ had had regional representatives for over twenty years now. Some of them have done remarkable things, some nothing at all, and some something in between. I originally intended my first wife to manage the list of regional reps, but as history shows, that never happened.

Ever since Alison and I split up I have been intending to ask someone else to take over the job, and finally a few months ago I got around to it. Ronan Coghlan has agreed to take over the onerous task, and has come up with a list of suggested roles for regional representatives, which I post here for public discussion.

[1] In the event of a reported sighting of a mystery animal in the representative's area, all possible data should be gathered and forwarded to CFZ. Likewise, news of further developments should be sent on as they occur.

[2] Representatives should try to discover if there were any sightings or other anomalous events in their areas in the past, but should only send on stories of UFOs or ghosts if they consider them important, as otherwise their is the danger of CFZ being swamped.

[3] Representatives should, if possible, look into local folklore to discover if stories of anomalous events in the area occur. Liaisons should be initiated with the Bird, Butterfly and Conservation Officer in their areas where possible. They should, in addition, try to gather an archive of Fortean zoological material from their local studies libraries.

[4] Representatives should initiate liaisons with groups dealing with anomalies and nature in the area, provided they consider them and their personnel suitable.

[5] Representatives should have the option of offering sales of books to local bookshops. However, some might find this distasteful and so this should not be regarded as an actual representative's duty.

In Search of Bowheads and Narwhals

By John Brodie-Good

An expedition voyage on the *Akademik Ioffe* from Iqaluit to Resolute, Baffin Island (August 2016)

Ever since I was a small boy I had marvelled at the legendary three species of high Arctic whales, beluga, bowhead and narwhal. All three are only found in the far north, and their lives are very dependant on sea ice. Belugas (the small, white whale or sea canary) are now being seen by most people who visit Svalbard during the summer expedition cruising season (June to early September). This is a pleasing change from ten to twenty years ago when they were very rarely seen there. Svalbard is the most accessible and 'cheapest' part of the true Arctic to get to from the UK. They occur in Arctic Canada too, most famously in Cunningham Inlet in 'spring'.

Bowhead whales were one of the first large whales extensively hunted by man, their numbers seemingly slowly recovering, despite very limited 'aboriginal' hunting which still continues to this day. Bowheads are astonishingly well adapted to their year round icy home. They have the thickest blubber of any whale, the largest mouth of any animal (one third of their entire length is head) and can break ice up to two feet thick with the top of their heads! They have no dorsal fin, which would get in the way of 'ice-breaking'. In addition, they have recently been discovered to live for over 200 years.

The unicorn of mythology could only have come from the narwhal. I have read claims that the existence of these unique animals was 'suppressed' in past years to lead more credence to the 'magic horse' stories. Long and sleek, the tusks on older males can exceed six feet in length, occasionally animals have two tusks (for stereo reception?). Like belugas, they are usually found in quite large groups; unlike belugas which are often friendly and very curious about us, narwhal have always been known to be notoriously shy, and will usually not tolerate any kind of ship or even outboard engine noise at all.

With the perceived intelligence of cetaceans, it would seem almost impossible not to believe that inter-animal communication, including from generation to generation, keeps both these species apparent fear of man alive today, and for good reason.

Many texts show narwhal as occurring in Svalbard, if they are still there, no one is seeing them. (A polar bear was however, seen and photographed dragging one out of the sea in 2015!). Bowheads can sometimes be seen in the spring in the North Atlantic, on the edges of the East Greenland ice-pack, or the southern edge of the ice-pack above Spitsbergen (assuming an extensive summer melt, eg 2016). You need a lot of luck though, no guarantee of seeing them at all. Pleasingly, over 80 bowheads were seen on one day in 2015, the highest number recorded since the 17th century in the North Atlantic. South West Greenland suffered a particularly heavy ice winter prior; these may have been those animals displaced? The usual, occasional sightings up here, would be two to four individuals together. Perhaps more incredibly, there are two recent UK records of bowhead, both thought to be the same young animal, both in Cornwall this and last year. Add an adult narwhal found 20kms up a river in Belgium this year (dead), you do wonder how fast the Arctic is changing.

So we flew north at the end of July to join the expedition ship *Ioffe* in Iqaluit (Frosbisher Bay), at the south east corner of Baffin Island, and slowly worked our way up the eastern coast and through the Parry Strait for our last landings on Devon Island, at the eastern end of the North West Passage. *Ioffe* was purpose built (with her sister ship *Vavlilov*) for scientific research for an Russian Oceanographic Institute and have a virtually silent sonic footprint. It makes them amongst the best for 'whale-watching' as a bonus. I have seen a female humpback using *Ioffe* as a 'back-scratcher' in Antarctica before now…..Satellite ice-maps showed the presence of very heavy sea-ice, hemming in the coast ahead of us. This included the entrance to Isabella Bay, a few days north of us. A bowhead sanctuary, where good numbers come to mate each summer. We had permission to go in, but could we get there? ……

What follows is extracts from my full 'voyage 'report'.

1st August
Frobisher Bay (Baffin Island)

The province of Nunavut, 'our land', is the size of Europe and inhabited by only around 35,000 people. These are predominately native Inuits, who only a generation or two ago were still being born in igloos, and lived in fur-walled tents in the summer, but are now of course very much in the 21st century. Modern housing, supermarkets, fully loaded 4x4s, speedboats, skidoos and high powered rifles. They currently also fully claim their traditional harvesting of all animals and birds rights,

ᐅᖃ ᐅᖅᑐᖅ
Nunavut

which in their present world of seemingly rapidly melting ice, I feel needs reviewing.

Our expedition leader was Boris Wise, supported by an excellent team, including birder/naturalist Jacque Sirois. Jacque is an immensely likeable Canadian, with a French accent, and a twinkle or ten in his eyes. He birds his local patch on Vancouver Island by kayak from his house, the patch being a group of small islands offshore. Jacque was a mine of interesting background information and had sailed these waters a number of times before. We also had an Inuit 'hunter' onboard, Ted, also extremely likable and interesting.

2nd August

At 1815 Boris called us downstairs for a briefing on the voyage ahead; sounded a bit ominous? Both Bill and I had been looking at the Canadian ice-charts online prior to departure from the UK. They clearly showed the southern part of eastern Baffin Island hemmed in by some serious sea-ice. The big question was, when, and how quickly was this going to break up? Boris to be fair, showed a series very recent maps, including today's. It was immediately clear that plans for the first few days were going to have to change. Whilst there were areas of less ice, these would have seriously slowed us down, impacting further on the planned itinerary later on. It had been decided to effectively skirt the outside of the sea ice, and see where we could get into the coast, further north. It sounded like one, or even two settlement landings were under threat as a result. This was no skin off our noses but Isabella Bay (the bowhead sanctuary) was also currently blocked in. I realised we would have to work even harder, and try and find some sooner ideally.

3rd August
Davis Strait (Lady Franklin Island area / Ice Pack cruising)

In the afternoon, a stop was made and a short zodiac cruise into the surrounding ice was offered. I declined, I had already spoken to Ted who had confirmed we were already in prime bowhead and narwhal country, so Sarah-Jane and I stayed on the ship and continued scanning from it's higher elevation.

5th August
Sunshine Fjord & Akpait National Wildlife Area, Baffin Island.

During the night, the Captain had sailed through a 'gap' in the heavy sea ice so we could attempt some landings. We sailed through the narrow entrance to Sunshine Fjord in the morning, which just straddles the Arctic Circle. We stayed on ship scanning for narwhal. This type of steep-sided fjord are their classic summer home. In spite of looking so fantastic in the books, they are often very unobtrusive, and people who have seen them before have often said they were 'underwhelmed'. They don't like ships or zodiacs which makes them even harder to see! Still, nothing like a challenge. I spent the morning slowly scanning the sides of the fjord with my scope. I was struck by how large it was, and how distant birds looked. This was not going to be easy. With everyone safely back on board, we hauled anchor and starting heading back out into the ocean. I stayed out on the bridge scanning. A few kittiwakes were flying across the entrance of the fjord. A local, very cold wind blew up, and as I was waiting for a call on the radio for the main course of lunch, Boris suddenly pulled open the bridge door and said 'whales, probably bowheads!' I ran through the bridge and managed to see three animals blowing right next to the cliffs on the south side, the blows turning into V's. I called down on the radio and grabbed my scope from the other side. Boris put out an announcement on the PA and surprisingly only three people came up behind the rest of our crew. Another set of blows (vertical only) and this time some black body too. The Captain slowed down but said no charts, he wasn't able to change course, and we were soon in the fogbank offshore. I had just learnt that this was also the summer home of bowheads, which are often found around the immediate entrance of fjords such as this. They too are not keen on ships either I was told….Frustrating, but at least a glimpse of high Arctic whale number two on my list of targets. (Noteworthy that the sister ship *Vavlilov* had the same three animals at the same spot, 24 hours later.)

The fog was fairly quickly blown away by an offshore breeze, we cruised north along the spectacular coastline, miraged ice bergs in the distance, and for the first time on this voyage, white caps, the fulmars finally being able to glide and arc as they were born to. Numbers of Brunnich's were now rapidly building as we headed towards a colony of over 100,000 pairs. We arrived late afternoon and the zodiacs took us on a short cruise beneath the towering cliffs, splattered with centuries of guano. Kittiwakes occupied the lower levels with the auks higher up. The tops of the cliffs had been eroded into spires by the winds, above which the guillemots flew like midges. On some small pieces of ice, small numbers of guillemots sat, looking like penguins down south. Quite a few seemed to be feeding just under the cliffs,

coming up with small fish in their bills; Arctic cod. Suddenly a shout on the radios, another pair of bowheads, in front of the ship, I turned just in time to see a blow! We cruised over to the area where they were heading and waited. Ann & Sarah-Jane saw a fluke which was sadly not followed up by our driver and that was it. We cruised back to the ship, which then turned back south. Boris informed us that the latest ice chart meant we couldn't continue north along the coast but had to swing out east and then head north, missing Isabella Bay. I hadn't woken up this morning thinking I would see bowhead today, and I went to bed knowing that I had only partially seen a bowhead. Missing out their main breeding location in this part of the Arctic didn't make me feel much better.

7th August
Baffin Bay and Gibbs Fjord

We spent the afternoon cruising slowly along the 15 miles of this spectacular fjord, although most of the upper half of the sides were shrouded in cloud still. Occasionally the sun broke through, highlighting the different colours of the melt waters from the glaciers to the seawater, with sharp delineations clearly visible between the two. The colder ice water was effectively floating on the saltier water below. We enjoyed glaciers, ice cliffs, scree slopes and hanging glaciers whilst we constantly scanned the fjord edges. I was again quietly concerned at the scale of this landscape, and the distance we were from the edges, where our quarry was likely to be. All too soon we had turned around and were now steaming much faster back out again. Boris had hoped we might still be able to turn back south towards Isabella Bay but the sea ice still held so we were going to head north. However, we had a day in hand so we would try a potential 'narwhal' fjord instead; that sounded much brighter at least.

8th August
Buchan Gulf & Icy Arm

We sailed into fjord mouth, and lots of fulmars (a colony is nearby) wheeled around us briefly. After a hurried breakfast, back up on the bridge wings. I started scope scanning below the cliffs on north side, I was feeling low again, it was another big area, surely no chance. I just finished a scan at a little headland when I saw little puffy blows and then grey shapes appear in the water below them. I knew, I just knew. I screamed 'narwhal' into the radio and into Boris's ear who was just inside the bridge door. The ship slowed and stopped and soon everyone was out on deck. We had come to a stop in front of a small river flowing into the bay, black cliffs to the left, and then a picture

perfect small glacier flowing into the bay further left again, with mountains above all of it. We only had four scopes between us but with Boris and others' assistance we worked out that there seemed to be at least two groups of animals. An American lady appeared with an expensive Austrian scope and tripod but didn't seem to know how to set it up. Jacque quickly sprang into action, and started watching the narwhals through it. He turned to her and quipped, "are you married?" The initial small group were now in front of the river mouth, and smaller groups were spotted swimming right along the edge under the black cliff, which were actually closer. The photographers, even those with the biggest lens, didn't stand a chance, way too far away. I wanted to see and watch them. Through bins they were small, through the scope you could get decent views. You picked them up by their quick, rapid blows, slight splash and then rolling backs. Even so, it took a good hour's worth of views to get an even fairly complete picture, little showed for long. At one point though, I got even luckier, a white-tipped black tusk came out of the sea in my field of view followed by the beautiful jet black body with the pure white mottling of an adult male. The smaller females were brownish but almost equally beautiful as their backs broke the dull grey water's surface and the younger animals seemed greyish. The lack of dorsal fin was apparent, and the ridge along the back was visible on several occasions too, as was two views of their rather small flukes. As the most left-hand group approached the glacier front Sarah–Jane said "Polar bear". To cap the spectacle, a bear had appeared to get up and start walking across in front of the glacier. At one point, I briefly had the glacier, a polar bear and two narwhals in my bins' field of view! The bear entered the water and swam for a bit. He then exited and climbed up an adjacent hillside. When we sailed past back out in the evening, he could be seen stretched out laying down, dozing. What a morning, what an animal narwhal is; they deserve a lot more than to be hunted.

We sailed on deeper inland to Icy Arm, where Boris planned a landing, a new site. We encountered the same, or different, narwhals again, swimming very close to shore, in the same direction. They soon turned, the views not as good as the morning. Boris mentioned if we were very lucky, once we were at anchor and the main engines off, they may come back. He admitted this had happened to him once, they stood on land with narwhals in the water yards away from them, and for as long as they wanted! Needless to say, it didn't happen today. But we were still very happy. We enjoyed a short walk ashore in the afternoon sunshine, examining the various flowers, seeing three species of butterfly (!) and the infamous Arctic bee, a bumblebee that can survive this far north. The scenery was spectacular, the sheer-sided cliffs and mountains being still capped by ice above. It was actually the first time we had been ashore, good to stretch the legs, Boris and Ted stood guard with rather serious looking shotguns. As we sailed back up the narrow arm we spotted the narwhals again, but they disappeared rapidly ahead of us into a bay to the left. We reached the open sea and headed north again, the stunning mountains and glaciers of Baffin Island on our port side in the evening sunshine.

9th August
Pond Inlet & Eclipse Sound

Another blue-skied, no wind day dawned, we went ashore after breakfast, this community of around 1500 people being the gateway to the far north. Rose, our local guide, was standing on the beach dressed in the most beautiful traditional sealskin smock, with a baby in the hood. We wandered slowly along the beach, seeing ravens and a few Lapland buntings (females and juveniles only) plus four rather cute husky pups and three Airbus A380s flying way high above us, probably heading for San Francisco from Europe. We walked along a road towards the community centre, where a show about their former traditional lives was being put on for us. Ann had discovered the narwhal meat store on the way and had been brave enough to go in, a small building with freezers inside. The reality of 'harvesting the local wildlife' was now staring us in the face. We assembled in the purpose built hall and I spotted a box of narwhal badges on a counter. Perfect 'medals' for members of the Narwhal Club! Before the excellent little performance began, a few words were said and two things became clear. The first was they, the locals, were worried about the warmer summers and melting ice. The lady pointed out that the group of mountains across the water from us, which had a few patches of snow on them, was covered by glaciers only 20 years ago! Secondly, they seemed acutely aware that many in the outside world had issues with their continued hunting lifestyle. In fact, I had been hearing that the regional council of elders seemed to be trying to ban all shipping from Eclipse Sound and both Milne and Navy Board Inlets, to 'protect their traditional hunting grounds'. Or perhaps to stop 'outsiders' witnessing what goes on? Subsistence hunting in canoes with hand help harpoons is one thing, mass slaughter using rifles, speedboats, snowmobiles and sometimes fuelled by alcohol is frankly just a turkey-shoot. With a Co-Op supermarket stocked with many of the produce and products we can all buy at home in the settlement, the hunting argument becomes weaker still.

There is no protection for narwhals and belugas at all. Bowheads are limited to five animals a year. It has recently been discovered that some bowheads are over 200 years old. An animal like that, which only just survived the hunting slaughter of the white man during the previous two centuries as a species, I think has earned its right to complete peace. Most of the animals and birds we saw during the voyage seemed wary of man, not surprising really. The government estimate the annual 'harvest' is worth C$40M per year, and that the replacement 'food value' would be C$5M p.a. With the province's income from mining and fishing exports shooting up, the C$5m would not even be noticed. I feel the time is coming for the Inuit to sit down with outsiders and discuss the future of their region's wildlife. At the very least, serious moderation is required? Perhaps they could follow examples set in other parts of the world and become wildlife guides, and not hunters? The show included demonstrations of their unique Northern sports, dancing and throat-singing. Some local art was for sale, we could not buy much of it of course (banned animal products) but I did purchase some rather nice polar bear cards.

The Co-Op was somewhat surreal. As you walked in, one of the ladies from the information desk was standing by the desk, to greet everyone, but she was holding a narwhal tusk in each hand! A ship was in, and they knew it. One of the now whitish polished tusks was just over 6 feet long. I briefly held it, it was pretty heavy, C$1300. Thankfully, no one from our vessel was tempted enough (only the Canadians could have purchased them legally).

We spent the afternoon cruising west in the middle of Eclipse Sound, with the vast nature reserve of Bylot Island off our starboard side. Being in the middle of this quite large body of water meant kittiwakes were the predominate bird species. Pleasingly, a good number of long-tailed skuas were also present, totalling about 50 in total with a close group of six birds at one point. But for us, Bylot Island was simply too far away. I bumped into Boris downstairs on a loo break and said 'any chance of moving into the coast more please?', I could see the bridge had accurate charts. He said he would see what he could do and shortly appeared up on the bridge and spoke to the watch officer. We turned north and started cruising quite a bit closer now. After only a few minutes Boris popped his head out and pointed out a group of 20 adult harp seals in the water, right under our noses. A few minutes later he was out again, 'Bowhead ahead'.

We had been out on deck for hours with little joy, he comes up to the bridge and spots the good stuff within minutes! I locked on to the blowing whale in the distance before it dived properly, it was off a headland ahead of us. I glanced at my watch and noted the time. The ship slowed down and we waited, scanning all around. After nearly 25 minutes it came up again, this time parallel to us but heading astern. I got my scope on it straight away and started soaking it in. The 'two humps' effect was now clearly visible when the animal surfaced, I was also looking straight into the animals blowhole too, 'elephantine' is a good description. As this could have been our last bowhead (and it was), the order

was given to follow the whale for a while. We turned back and began slowly moving again, the Captain perfectly staying away and behind where we thought the animal was heading. After about another 20 minutes it came back up to multiply breathe again before sounding. Annoyingly, in spite of what appeared to be a few big rolls on the surface, it didn't show its flukes at all. After our 4th view we turned back on course and left it in peace. Even when you are watching them, these Arctic whales don't give up their

Copyright Dick Filby

secrets very readily! I guess it's part of their mystery.

August 10th
Low Point & Navy Board Inlet

We also enjoyed decent scope views of a ringed seal hauled out on a small piece of ice. Both Ann and I separately saw what we thought to be a large dark shape in the water looking towards Bylot, both of us thought the word bowhead but nothing further was seen.

August 13th
Resolute & Edmonton

As we got airborne the pilot banked sharply to the left, the *Ioffe* at anchor in the bay below us. The incoming passengers were heading further west including to Cunningham Inlet to look for belugas. We envied them, although ultimately had no reason to complain after our wonderful voyage. As we climbed towards altitude we could see into the North West Passage itself, on the right hand side of the plane, large amounts of sea ice including some really big floes.....it looked amazing...maybe one day...........

We over-nighted at a very comfortable airport hotel at Edmonton and Air Canada safely delivered us back to Heathrow the following morning after that. We had just been to one of the least visited parts of the world, and we had seen the unicorn. They do really exist.

Optical gear: Leica 10 x32 binoculars & 20 – 50 x 60 fieldscope.

Thanks to One Ocean Expeditions, the Captain and crew of the *Akademik Ioffe*, Boris Wise and his excellent team of staff.

For the full voyage report, including multiple polar bear encounters, other whales and dolphins please visit

http://www.wildwings.co.uk/app-holidays/ baffin-island-canadian-high-arctic-new

Species Lists
Baffin Island Voyage 1st – 13th August
Mammals
Bowhead
Fin Whale
Northern Bottlenose Whale
Long-finned Pilot Whale
Narwhal
Sowerby's Beaked Whale
White-beaked Dolphin
Bearded Seal
Hooded Seal

Ringed Seal
Harbour (Common) Seal
Seal sp
Harp Seal
Polar Bear
Arctic Fox
Arctic Hare
American Brown Lemming
Musk Ox
Birds
Eider
King Eider
Long-tailed Duck
Greater Snow Geese
Brent Goose
Canada Goose
Red Throated Diver
Great Northern Diver
Fulmar – nominate white birds plus blue and brown morphs.
Dunlin
Purple Sandpiper
Semi-palmated Plover
Pomarine Skua
Long-tailed Skua
Arctic Skua

Glaucous Gull
American Herring Gull
Thayer's Gull
Kittiwake
Rough-legged Hawk
Gyr
Peregrine
Guillemot
Razorbill
Brunnich's Guillemot
Black Guillemot
Little Auk
Puffin
Grey Phalarope
Red-necked Phalarope
Raven
Snow Bunting
Lapland Bunting
Buff-bellied Pipit
Other wildlife
Arctic Fritillary
American Copper
Hecla Sulphur
Arctic Bumble Bee
Arctic Cod

The Hissing of Summer Lawns

Letters

The editor and his compadres welcome letters for publication on all subjects covered by this magazine. However, we would like to stress that neither this magazine, or the CFZ are responsible for opinions expressed, which are purely those of the letter writer.

Dear Readers

SURVIVAL OF THE MAMMOTH INTO LATE 18TH CENTURY NORTH AMERICA

I recently came across a newspaper story titled :

"THE MAMMOTH or BIG BUFFALO. [From Jefferson`s Notes of Virginia, a work not yet published] In the *Pennsylvania Packet* of June 13th, 1787. :

(It was the contention of the author that the elephant and the mammoth were different animals, which of course we today know to be true, but in those days it was still a matter of debate.)

"Our quadrupeds have been mostly described by Linnaeus and Mon. de Buffon. Of these the Mammoth, or Big Buffalo, as called by the Indians, must certainly have been the largest. Their tradition is he was carnivorous and still exists in the northern parts of America. A delegation of warriors from the Delaware tribe having visited the governor of Virginia, during the present revolution,, on matters of business, after these had been discussed and settled in council, the governor asked them some questions relative to their country, and among others, what they knew or had heard about the animal whose bones were found at the Saltlick on the Ohio..."

[The text then describes a fanciful legend of how the Native Americans version of God punished the Big Buffalo for its destruction of other large mammals and removed it to over the Great Lakes where it still lived in the 1780s]

A Mr Stanley, taken prisoner by the Indians, near the mouth of the Tanissee, relates, that after being transferred through several tribes, from one to another, he was at length carried over the mountains west of the Missouri, to a river which runs west-wardly, that these bones abounded there; and that the natives described to him the animal to which they belonged as still existing in the northern parts of their country; from which description he judged to be an elephant...I will not avail myself of the authority of the celebrated anatomist (Hunter), who, from an examination of the form and structure of the tusks, has declared they were essentially different from those of the elephant; because another anatomist (D`Aubenton), equally celebrated, has declared, on a like examination, that they are precisely the same...

[The text then goes in to a long discourse on the differences between the elephant and the mammoth.]

..But no bones of the Mammoth, as I have before observed, have been ever found further south than the salines of the Houston, and they have been found as far north as the Arctic circle.. "

Best Wishes

Richard Muirhead
Macclesfield

Miles of Aisles

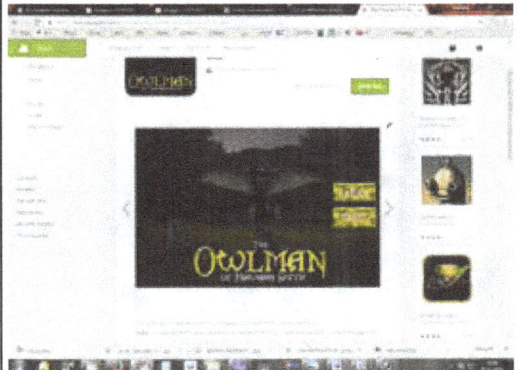

I was just looking for useful apps for my new Garmin GPS device when I came across this:

https://play.google.com/store/apps/details?id=com.steampunk.owlman

I didn't know if you were aware of it. I didn't buy it as looking for useful stuff not games.

Steve Jones.
Wakefield

Reviews

The first scientific evidence
on the survival of apemen
into modern times

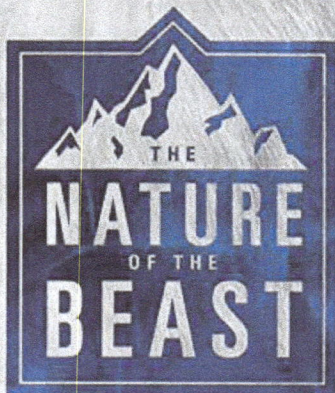

THE
NATURE
OF THE
BEAST

BRYAN SYKES

Professor of Human Genetics at
Oxford University and author of the international
bestseller *The Seven Daughters of Eve*

The Nature of the Beast
Professor Bryan Sykes
ISBN-9781444791259
Coronet 2015

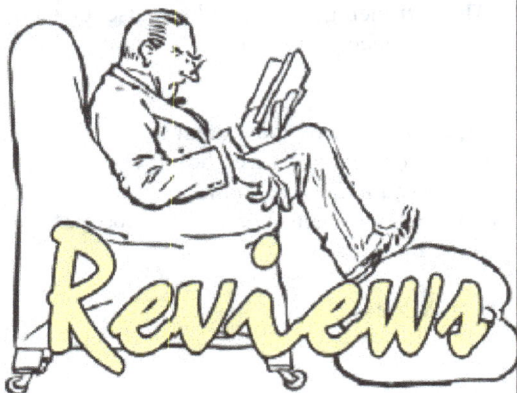

Many books have been written on mystery hominids and hominins over the years, including a number of classic titles. Ivan T Sanderson's *Abominable Snowmen: A Legend Come to Life*, Ralph Izzard's *Abominable Snowman Adventure* and Janet and Colin Bord's *Bigfoot Casebook* spring to mind.

The Nature of the Beast is something quite different to anything that has come before. Not only is it the best book ever written on the subject, it is the most important. The book is authored by Professor of Human Genetics of Oxford University, Bryan Sykes, who is one of the world's leading geneticists. For a scientist of such standing, to put his head over the parapet in such a contentious subject takes a lot of guts, but the end result is worth it.

Together with Michel Sartori, the Director of the Museum of Zoology in Lausanne, Switzerland, Professor Sykes instigated The Oxford-Lausanne Collateral Hominid Project. The idea was to bring hard science into the search for man-like monsters. Sykes and Satori invited people to send them supposed hair samples from unknown primates such as the yeti, sasquatch, almasty and orang pendek.

Sykes has researched human origins for over twenty-five years via the study of mitochondrial DNA. This is inherited from the maternal line and is generally the best preserved DNA. Mitochondria are found between the cell wall and nucleus of each cell, and release energy; ergo they are relatively abundant. Simply put, Sykes has perfected a technique that examines a DNA segment called 12S RNA, part of a gene that helps mitochondria assemble the enzymes required for aerobic metabolism. This sequence is known for all known species of mammal. Hence there could be no confusion in any sample sent to the Project; they would be from one of the known species or from something new. A hypothetical new species could have its place on the genetic tree revealed by its closeness to other species. This meant that human contamination could be avoided. Even Neanderthal 12S RNA differs from modern man.

The book is the result of the analysis of thirty samples sent to the project and Sykes' own travels and personal researches. Sykes finds himself in the wilds of the American North East listening to what may or may not be an unseen sasquatch banging the walls of tunnel beneath a tree. He ventures to Russia, home of the Snowman Commission, an official government backed project to hunt hominins in the former Soviet Union. Set up in the 1950s, it only lasted three years but has recently been re-formed. The Russian scientists he meets are all confirmed believers and oddly seem to think it important that Bryan only published positive results from his work, and not negative. At the museum in Lausanne he examines the massive wealth of information, clippings, letters and writing bequeathed by Dr Bernard Heuvalmans, the man widely accepted to have created the discipline of cryptozoology in its modern form. There is a highly interesting chapter on the Minnesota Iceman debacle were Heuvalmans and his friend, the Scottish zoologist Ivan T Sanderson, seem to have been duped by a clever fake, a faux ape-man in a block of ice. Correspondence between the two does not show the 'father of cryptozoology' in the best of lights.

The professor meets modern explorers and cryptozoologists such as Jon Downes and myself at the Centre for Fortean Zoology, Loren Coleman, Dr Jeff Meldrum, and the mountaineer Reinhold Messner.

The hair samples were collected from all across the world and sent in to the Project. Each and every one turned out to be from known species.

Bears, horses, humans and goats were among the creatures found to be the former owners of the hairs.

The book is completely honest and open in its treatment of the subject. The professor is clear in pointing out that just because these samples turned out to be from known species this does not mean that anomalous primates do not exist. Sykes criticizes some cryptozoologists for not being rigorous enough and going through the right channels in the analysis of samples. He equally lambasts some scientists for rejecting the notion of large unknown creatures out of hand.

Despite the negative results of the hair analysis the professor's personal view is more positive.

"Funnily enough, even though there were no anomalous primates in among the hairs I tested, I think my view has altered more to 'something out there' than the reverse. The change of heart comes from speaking to several people, some not even mentioned in The Nature of the Beast, who have nothing to gain but who have seen things in good light while in the company of other witnesses, that are hard to explain otherwise. To automatically reject these accounts is just as blinkered as accepting that every broken branch has been snapped or twisted by a sasquatch."

In the book's stunning postscript, the professor's optimism may just have been proven well-founded. This section of the book astounded me so much I had to re-read it in order to make sure I had not been mistaken. The professor may now be on the edge of a jaw-dropping discovery.

Whilst in Russia he was able to secure a tooth from a remarkable skull once owned by the Darwin Museum in Moscow, but subsequently sold to a private collector. The skull was from a man named Khwit. Khwit, (below) who died in the early 1950s, was said to be one of several hybrid children born from an almasty mother and a human farther. The almasty are said to be large, powerfully built, hair-covered, wild people reported from the mountains of Russia and the former USSR. They are more man-like than the yeti, but clearly not modern humans.

In the 1850s a female almasty was captured in a forested region of what is now Abkhazia in the Caucasus. Tall, muscular with an ape-like face and covered in reddish black hair, she was named Zana. Zana was taken to the farm of a local nobleman in the village of T'khina. She finally became tame and could do menial tasks around the farm with her immense strength. She never learned speech, but made inarticulate noises. Zana became the mother of a number of

hybrid children via a village man, Edgi Genaba. The first two children died after their mother tried to wash them in a cold river. Subsequent children were taken from her and raised by villagers. The two boys, Dzhanda and Khwit Genaba (born 1878 and 1884), and the two girls, Kodzhanar and Gamasa Genaba (born 1880 and 1882), were assimilated into normal society, married, and had families of their own. Zana herself died in 1890. Russian researchers located the grave of Khwit and recovered his skull.

Extracting mitochondrial DNA from the tooth Professor Sykes found that it was 100% sub-Saharan African. This was confusing, as Zana was clearly no kind of modern human. Her behaviour and appearance seemed to be far more primitive than a Neanderthal. Further work showed the DNA was from an exceptionally ancient lineage from Western Africa and furthermore it may have been pre *Homo sapien.* Sykes thinks that this lineage may have left Africa over 150,000 years ago, before modern man. If he is correct then Zana may have been an unknown species of pre-human hominin, a species still lurking in the Caucasus and other areas today.

I was in recent e-mail contact with Professor Sykes. Like all good scientists he is proceeding with caution. He and another geneticist are still working on the sample. He has likened Zana's DNA to old fragments of old photographs that have filtered down through time. I await the results with baited breath.

The Nature of the Beast could easily have been a dry tome, but it is written in a highly entertaining and excusable manner yet without losing its scientific credibility; a very impressive feat. It is a must read for any cryptozoologist or Fortean. **RF**
10 /10

DIDN'T WE HAVE A WEIRD WEEKEND?

The third weekend in August in 2016 saw the 17th Weird Weekend, and it was a particularly poignant one because, although it may not be the last one that we ever hold here in Devon, it is certainly the last one which will be held in the current format. My announcement of this during my keynote speech on the Sunday afternoon did, I'm afraid, overshadow the rest of the event to a certain extent, but I would hate it to detract from what was one of the nicest events that we have ever held.

Once again it was held at the Small School, Hartland, North Devon, and as I have written elsewhere in this issue, I am sad to say that the school will be closing at the end of the current academic year. We are living in unpleasant times when Independent organisations such as the Small School and the CFZ find themselves in increasingly straightened circumstances, and – increasingly – are going to the wall.

However, as I have explained elsewhere, the Weird Weekend is not really a victim of the current socio-economic climate. Rather it is a victim of my own failing health. I was forty when the event began, and now I am only a few years short of sixty, and in a wheelchair much of the time.

My friend Alan Dearling was one of the speakers this year but he also wrote a review which was published in *Gonzo Weekly*. I have taken the liberty of printing some of it here:

"… There were moments that were straight out of Monty Python or a Ken Campbell show. Such as when the loveable and clownish, Richard Muirhead produced the book: 'Thought Transference in Birds' (1931) and Jon Downes gleefully cried out something like,

"My dear boy, give it to me. We must reprint it. What about the film rights?"
Some fun and serious bits

My personal favourite inputs were:

Jon Downes' master of ceremonies quips, comments on a cornucopia of subjects and especially his closing address on the future of Weird Weekends. Jon is larger than life, and as large as Moses, David Bowie and possibly God. See part of his closing epistle at: https:// www.youtube.com/watch? feature=player_embedded&v=bM5u7socDnY

Lars Thomas gave an amusing and factual talk on the Vikings and their monsters.
Engaging, humorous and informative - Steve Ignorant's incomparable performance show on the history of Mister Punch was just great. I've written about it separately. It was also very non-PC. You can see a bit of it at: https:// www.youtube.com/watch? feature=player_embedded&v=0cseVgOfXhs

Matthew Watkins is an extraordinary mathematician, and I found his subject matter – retrocausality (can the future come before the past) fascinating, but it did cause my little brain to hurt a bit. Easier on my brain cells was Shoshannah McCarthy's well-illustrated and proven dissection of the myths surrounding

vampire dogs.

Author, Julian Vayne, gave a veritable Cook's Tour, complete with useful visuals across a wide range of Chaos Magickians, including Aleister Crowley, Madame Blavatsky, Austin Osman Spare (one of my personal favourites – occultist and artist – the originator of the Sigil) and Peter J. Carroll.

And finally, I loved the droll Irishman, Ronan Coghlan, who reminded us of the spell-binding power of the story-teller. His input on the possible histories and myths of Robin Hood was tour-de-force. You can catch the start of his 'show' at:

https://www.youtube.com/watch?feature=player_embedded&v=UX3Ki2i0U2E
"

The full speaker list is as follows:

Steve Ignorant: The hidden history of Punch and Judy
Jackie Tonks: My encounter with Bigfoot
Lars Thomas: The Vikings and their Monsters
Richard Freeman: The Almasty
Steve Rider: Tales from the SUFOG archive
Glen Vaudrey: The Health and Safety Beast
Alan Dearling: Futurology and free cultural spaces around the world
Julian Vayne: Chaos Magick
Music from **Stargrace**
Richard Freeman: Tasmania 2016 Expedition Report
Richard Muirhead: Weird Fauna: Peculiar and Rarely Observed Animal Behaviour
Matthew Watkins: Retrocausality and other reverse-time phenomena
Shoshannah McCarthy: Vampire dogs
Ronan Coghlan: Robin Hood: Origin of the

legend
Jon Downes: Keynote Speech

As I have done for the last 17 years, I chose the speakers personally, not only because I thought that they would be good and the sort of thing that a Weird Weekend audience would like to witness, but because they were things I would like to see myself. Many of the speakers are personal friends of mine, and one is family, so it would be completely inappropriate to single any of them out for praise or censure (not that anybody deserves the latter). However, I would like to talk about one particular performer.

A long time ago, back when my father was dying, a friend of ours put on a benefit gig at a pub in Barnstaple. It features several local punk bands and she asked me if I could write a small piece for them on why punk bands and cryptozoology are relevant to each other. I wrote the following:

"To me, punk was and is as much about the attitude and the politics than the music. I was a punk the first time round, and the libertarian do-it-yourself ethic of the movement has influenced me - and the CFZ - ever since. Even the name of our journal comes from a track by Adam and the Ants from their 1980 album 'Dirk wears White Sox'. A quarter of a century ago it was my privilege to meet CRASS on several occasions. They were an Essex based anarchist collective who were not only responsible for a string of gloriously nasty albums but managed to sell over 2 million of them. Not only this, but they managed to do it all themselves without the backing of record companies or the music industry. When I visited them I was impressed how the whole family worked together to run the office, and how - despite the lack of professional involvement - they not only managed to get all their product assembled and sent to the

retailers, but they managed to be cheaper (and better) than anything coming out of the mainstream music industry.

Nearly fifteen Years ago I founded the CFZ. I did so for precisely the same reason that Penny Rimbaud and Steve Ignorant founded CRASS in 1977. The cryptozoological establishment was moribund, boring and - worse of all - becoming increasingly corrupt. It was time for a new broom. Everyone from the old guard sneered at me when I told them what I was gonna do. "You can't start a scientific organisation without the backing of either industry or one of the universities" they said, as they carried on planning research whose only real motive was self aggrandisement or to make rich people even richer.

I remembered the tumbledown Essex farmhouse full of punks tirelessly stuffing LPs into envelopes and grinned to myself. I went on with my plans and now we are the biggest cryptozoological research organisation in the world...and we are still not, nor will we ever be in bed with any multinational corporation or establishment university if it means that the integrity of our work is threatened.

Cryptozoology is the study of unknown animals, but what we do is more than just looking for new species. Knowledge should be free and available to all. All too often these days it isn't. Not only are we untainted by association with major establishment figures, but our research is available to EVERYONE - whether or not they are members of the CFZ. We are working towards building a full time Visitors Centre - a truly global community resource and a place where researchers from all over the world can meet and work together.

I haven't really thought about it before, but I guess that not only was I a punk the first time

round, but I still am!

ANARCHY   PEACE   AND   FREEDOM

JD"

Back in 1981, like so many other young people I was politicised by Crass. Whilst I enjoyed the records I was never particularly convinced by the politics of the Sex Pistols or The Clash. However, I had not found any social or political philosophy that made sense to me. But when I discovered Crass and there polemic of vegetarianism, anarchy, peace and freedom I was an immediate convert, and most of the time at least, I have followed this philosophy ever since. Certainly, the CFZ, the Weird Weekend, CFZ Press and other things that I have done over the years, have been based directly on the Crass business model. Or should I say the Crass anti-business model!

And so it was particularly poignant for me that the headline speaker at what may well be the last Weird Weekend promoted by me was none other that Steve Ignorant from the very band itself.

I don't know what's going to happen next. There are going to be Weird Weekends during 2017 in Cheshire promoted by Glen Vaudrey, and Copenhagen, promoted by Lars Thomas and Margit From. I suspect that there will be others over the years, and I wouldn't be surprised if Corinna and I do something of the kind at some point in the future.

But it is the end of this particular era, and the fact that the man who shaped the philosophy on which the whole CFZ edifice was built, was there with us at this historic event means more to me than I can possibly say.

THE WORLD'S WEIRDEST PUBLISHING GROUP

We publish a lot of books. Indeed, I think that we could quite easily claim to be the world's foremost publishers of books about Fortean Zoology and allied disciplines, and our Fortean Words imprint is doing a great job in producing books on other non-zoological esoterica. However, I feel that it would be unethical to review our own titles. So here, to end this edition of *Animals & Men*, is a brief look at the books we have put out since the last issue.

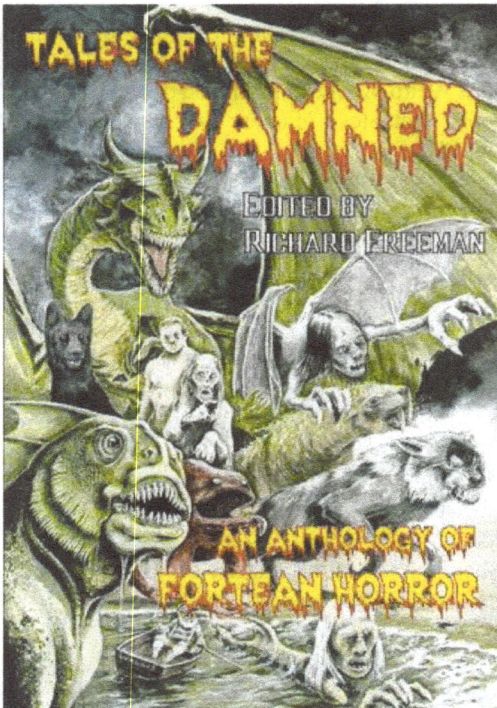

Here we have a book unique in the annals of horror literature. This is a book where the stories are penned, mostly, by Fortean researchers. For those who are unfamiliar with the term, 'Fortean' refers to the works of Charles Hoy Fort (August 6, 1874 - May 3, 1932).

Fort was an American writer and researcher into strange phenomena. Fort collected newspaper reports of sightings of strange creatures, weird light phenomena, falls of fish and other creatures, and poltergeist outbreaks to name but a few. Fort realized that mainstream science was acting somewhat like a fundamentalist religion and that any anomalous data that did not fit into the current scientific paradigm was simply swept under the carpet. Fort called this 'damned data'.

He published his findings in four books: *The*

Book of the Damned (1919), *New Lands* (1923), *Lo!* (1932) and *Wild Talents* (1932), all of which are still in print today. Fort has influenced modern research with magazines like Fortean Times, which has been recording weird happenings since 1973, and the Centre for Fortean Zoology, the world's only full time mystery animal investigation organization. The publishing wing of the latter is responsible for this book.

Here's Nessie!

a monstrous compendium from Loch Ness

Dr Karl P. N. Shuker

Nessie the Loch Ness monster (LNM) is not only the premier mystery beast of the United Kingdom, it also vies with the bigfoot or sasquatch as the most famous one anywhere in the world. Little wonder, therefore, that during his many years as a world-renowned cryptozoological researcher and writer, Dr Karl Shuker should have documented it and all manner of aspects relating to it in a wide range of publications

Now, however, for the very first time and in direct response to popular demand, all of Dr Shuker's most significant but previously disparate Nessie-themed writings have finally been brought together, and in expanded, updated form whenever possible too, to yield the present LNM compendium, covering a fascinating, extremely broad spectrum of pertinent topics. These include: a comprehensive review of the history and controversies associated with this exceedingly contentious aquatic cryptid; a diverse selection of the most – and least – plausible taxonomic identities that have been proposed for it; the closely-linked traditional Scottish folklore of kelpies and other water-horses; various Nessie-related hoaxes; an extensive survey of other Scottish freshwater loch monsters; reviews of Nessie-related material; a hitherto-unpublished LNM sighting from leading Nessie researcher and eyewitness Tim Dinsdale; Nessie in philately; the historic LNM conference staged by the International Society of Cryptozoology at Edinburgh's Royal Museum of Scotland in 1987; the enigmatic Pictish beast; a tribute in verse to Nessie; an annotated, YouTube-linked listing of Nessie-themed songs and music videos; and much more too!

Supplementing these varied subjects is an equally eclectic selection of illustrations – a dedicated Nessie gallery containing a dazzling array of spectacular full-colour LNM artwork, including a number of specially-commissioned, previously-unpublished examples, as well as a wide range of text images – plus a very comprehensive bibliography of non-fiction Nessie books, a listing of current LNM-themed websites, and a detailed index.

So without further ado, welcome to the sometimes decidedly weird yet always totally wonderful world of Nessie - the mystifying but ever-memorable monster of Loch Ness.

1st – 2nd April 2017

Rixton-with-Glazebrook Community Hall, Near Warrington, Cheshire

A must for anyone interested in Ghosts, Fortean, Fairies, UFO's, Folklore, Cryptozoology and other fascinating stuff

<u>Speakers:</u>

- Alan Murdie – Britain's most dangerous ghosts
- Glen Vaudrey - Zooform creatures
- Lee Walker - the Otterspool Timeslip & new Spring Heeled Jack sightings.
- Steve Jones – Haxey Hood and other village combat games
- Mick Walters – Mysterious Staffordshire
- Rob Whitehead - Out Of The Mouths Of Babes: Children's Encounters With UFOs
- Richard Freeman – Mongolian death worm
- Rob Gandy - Phantom hitchhikers on bikes
- Steve Mera - UFO's from pre biblical time to the present day
- Bob Fischer – Hobmen
- Jackie Tonks - Bigfoot in the pacific north west
- Tom Skelton – the phantom rabbit of crank

£15 per day or £20 for the two days

See: http://glenvaudrey.wix.com/weird-weekend-north

Animals & Men

Phantom Black Dogs in The Netherlands

PLUS: Tasmania 2017 Expedition Report, Phantom Kangaroos in North America, Tribute to the Bolam Beast, Return to Ape Canyon, plus news, reviews and more...

The Journal of the Centre for Fortean Zoology #60

Contents

Typeset by Jonathan Downes,
Cover and Layout by SPiderKaT for CFZ Communications
Using Microsoft Word 2000, Microsoft Publisher 2000, Adobe Photoshop CS.
First published in Great Britain by CFZ Press

CFZ Press, Myrtle Cottage, Woolsery, Bideford, North Devon, EX39 5QR

© CFZ MMXV

ISBN: 978-1-909488-31-1

Faculty of the Centre for Fortean Zoology

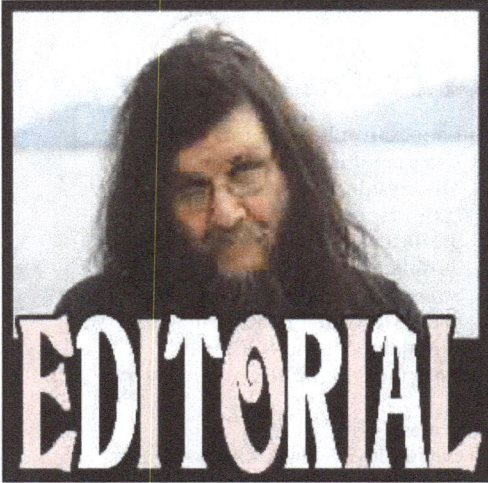

EDITORIAL

Dear Friends,

Welcome to the first issue of *Animals & Men* in our silver Jubilee. Yes, it was in the early spring of 1992 when my first wife and I were on tour with a band called Cockney Rebel, and we had reached the Scottish town of Inverness.

I had been interested in cryptozoology since I was about 8 years old, and so with several days off before the next gig, which was in Glasgow, an anabasis to the world's most famous cryptozoological location seemed an eminently sensible thing to do.

I was already thinking about how I could start a cryptozoological research organisation, but as my first wife was weary of my bright ideas which had a habit of costing more than I had

planned, I found myself, early one morning, sat on a rock on the shores of the lake trying to work out how I would justify the idea to her.

A quarter of a century later, and I haven't spoken to Alison in over 20 years, and it is unlikely that I shall ever do so again. However, despite the myriad unpleasantness that went down between us during our divorce, I would like to publicly thank her for supporting me during the first few years of the CFZ.

When I read Douglas Botting's biography of Gerald Durrell about 15 years ago, I was shocked to discover how, in the wake of his acrimonious

The Great Days of Zoology are not done!

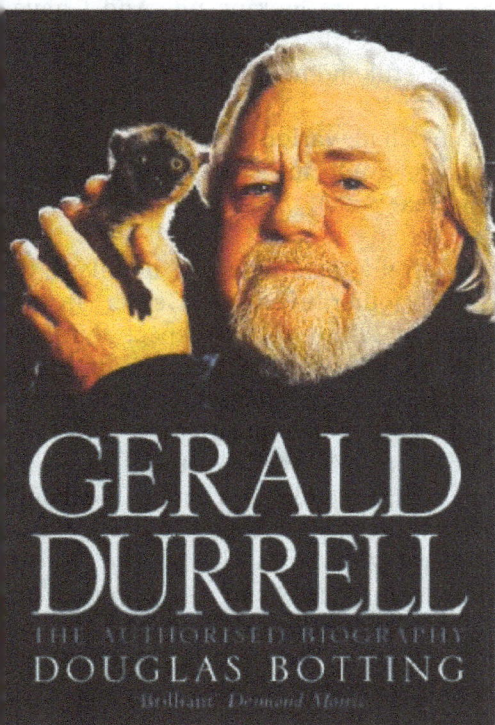

GERALD
DURRELL
THE AUTHORISED BIOGRAPHY
DOUGLAS BOTTING
Brilliant *Desmond Morris*

divorce, Durrell did his best to do a Stalinist rewrite of history and airbrush out the massive contribution that his first wife had made to his career.

Gerald Durrell is one of my greatest heroes, and peculiarly – he and I have quite a lot in common, in both good and bad ways. The difference between us, however, being that he was a giant amongst men, again, both in good and bad ways, whereas my unconscious emulation of him (for the third time) in both good and bad ways, is a mere shadow compared to what he had done. For example, I am only two years shy of 60, and, by the time he was my age, Durrell had not just started a world famous zoo, written a string of best selling books and formed one of the world's most important conservation organisation, but had drunk himself to the edge of epilepsy and various other things that are completely beyond my reach. However, there is one facet of his personality that I have no intention of trying to emulate.

We may have gone out different ways, but if it had not been for Alison's help all those years ago, the Centre for Fortean Zoology would have never been more than a pipedream.

We have had a couple of very difficult years. And I am aware that they have been difficult for everybody and not just those of us who are trying to run the world's largest cryptozoological research organisation. However, historically we have always been staffed by volunteers, and when the people who used to volunteer for organisations such as our own are forced – by increasing draconian legislation - to get jobs, often in the face of medical advice, just in order to survive, both the CFZ and society are up the creek.

And so, whereas once upon a time, there where four or five people working pretty much full time keeping the CFZ going, together with another dozen or so – each week – putting in significant amounts of time to help, nowadays there are no full-time workers, as even Corinna and I have been forced into taking on outside work in order to keep our heads above water.

But, the CFZ is evolving, and we are continuing to make new ground.

Already this year, we have sent our fourth expedition to Tasmania, and we have our second expedition to Russia planned for this summer. Richard Freeman writes:

"Back in 2008 the CFZ took a major expedition into the Caucasus Mountains in search of the almasty. Now we are returning. A group of researchers from England including Chris Clark, Dave Archer, Richard Freeman, Jon Hare and Jackie Tonks will be teaming up with Ukrainian researchers Grigory Panchenko and Anatoly Sarendenko in a new search for the hominin in Karbadino Balkaria.

Set for late July 2017 the trip will return to the area surveyed in 2008 as well as moving into fresh zones. We also hope to secure a tooth from the skull of Kwhit, a possible human/almasty hybrid who died in the 1950s."

And I'm happy to say that for the first day in what seems like many months, Peter the CFZ gardener has spent a day working hard trying to make the CFZ grounds look slightly less like a First World War battle field, and more like the wildlife garden it used to be.

So, things are looking up. And I have every hope that this is going to be a successful and happy year.

I have had various people ask me if there is going to be a Weird Weekend this year.

I'm afraid that there are no plans to hold another event in Devon at the moment, although, I suspect that we will be doing one at some point when my health problems have stabilised. However, in the mean time, there is the Weird Weekend North in Cheshire in April, and Weird Weekend Scandinavia in Copenhagen in June. Richard will be talking at the Weird Weekend North, and I shall also be talking at the Weird Weekend, but in Copenhagen. We shall both be talking at the ASSAP conference on the 21st of May in Bristol.

So things are looking up, and we can confidently predict that we will have a steady stream of interesting things to share with you this year.

Onwards and upwards!

Jon Downes
(Director, CFZ)

WEIRD WEEKEND North

1st – 2nd April 2017

Rixton-with-Glazebrook Community Hall, Near Warrington, Cheshire

A must for anyone interested in Ghosts, Fortean, Fairies, UFO's, Folklore, Cryptozoology and other fascinating stuff

Speakers:

- Alan Murdie – Britain's most dangerous ghosts
- Glen Vaudrey - Zooform creatures
- Lee Walker - the Otterspool Timeslip & new Spring Heeled Jack sightings.
- Steve Jones – Haxey Hood and other village combat games
- Mick Walters – Mysterious Staffordshire
- Rob Whitehead - Out Of The Mouths Of Babes: Children's Encounters With UFOs
- Richard Freeman – Mongolian death worm
- Rob Gandy - Phantom hitchhikers on bikes
- Steve Mera - UFO's from pre biblical time to the present day
- Bob Fischer – Hobmen
- Jackie Tonks - Bigfoot in the pacific north west
- Tom Skelton – the phantom rabbit of crank

£15 per day or £20 for the two days

See: http://glenvaudrey.wix.com/weird-weekend-north

A LEGAL MATTER

Newsfile

New & Rediscovered

The Funky Gibbon

A gibbon living in the tropical forests of south west China is a new species of primate, scientists have concluded.

The animal has been studied for some time, but new research confirms it is different from all other gibbons.

It has been named the Skywalker hoolock gibbon - partly because the Chinese characters of its scientific name mean "Heaven's movement" but also because the scientists are fans of Star Wars.

The study is published in the American *Journal of Primatology*.

Dr Sam Turvey, from the Zoological Society of London, who was part of the team studying the apes, told BBC News:

"In this area, so many species have declined or gone extinct because of habitat loss, hunting and general human overpopulation.

"So it's an absolute privilege to see something as special and as rare as a gibbon in a canopy in a Chinese rainforest, and especially when it turns out that the gibbons are actually a new species previously unrecognised by science."

SOURCE: http://www.bbc.co.uk/news/science-environment-38576819

Dinky Dryas

The Lomami National Park in the Democratic Republic of Congo (DRC) in Central Africa is now home to a new population of the Dryas monkey. Originally believed to inhabit only one site on the planet in the Congo basin, this colourful and beguiling animal is about the size of a house cat.

Field teams from the Lukuru Foundation TL2 Project discovered it near the border of the Lomami National Park when they noticed a dead monkey with a local hunter. They later confirmed it to be a Dryas monkey, known locally as Inoko. First discovered in 1932 and believed to be nearing extinction due to its small population size and unregulated hunting, this species has perplexed scientists for decades because of its elusive nature.

The Dryas monkey (*Cercopithecus dryas*), also known as Salonga monkey or ekele, is a little-known species of guenon found only in the Congo Basin, restricted to the left bank of the Congo River. It is now established that the animals that had been classified as *Cercopithecus salongo* (the common name being Zaire Diana monkey) were in fact Dryas monkeys. Some older sources treat the Dryas monkey as a subspecies of the Diana monkey and classify it as *C. diana dryas*, but it is geographically isolated from any known Diana monkey population.

SOURCES:
- https://www.sciencedaily.com/releases/2017/02/170201092639.htm
- Colyn, M.; Gautier-Hion, A.; Vanden Audenaerde, D. T. (1991). "Cercopithecus dryas Schwarz 1932 and C. salongo Vanden Audenaerde, Thys 1977 are the same species with an age-related coat pattern". Folia Primatologica. 56 (56): 167–170. doi:10.1159/000156543.

New Squirrel in Italy

Although it is unusual, new mammal species are even found in Europe on occasion. The latest of these is a squirrel. Thanks to a detailed genetic study, morphological and ecological, a team of Italian researchers, coordinated by the University of Insubria, found that the squirrel populations present in Calabria and Basilicata - which had since 1900 been recognized as peculiar "to the point of being regarded as a subspecies of the European red squirrel' belong to all effects to a new species of squirrel. The southern squirrel, (*Sciurus meridionalis*), (Photograph by Anthony Mancuso) is endemic to the forests of the regions of Calabria and Basilicata, in the south

of the Italian Peninsula and is a "close relative" of the European red squirrel, (*Sciurus vulgaris*) that is present throughout the rest of Italy, with the exception of Sicily and Sardinia. This new species has a characteristic black colour with a white belly, unlike the European Common squirrel that has a colour that can range from orange-red to dark brown.

SOURCES

- Wauters, Lucas A.; Giovanni Amori, Gaetano Aloise, Spartaco Gippoliti, Paolo Agnelli, Andrea Galimberti, Maurizio Casiraghi, Damiano Preatoni, Adriano Martinoli (2017). "New endemic mammal species for Europe: Sciurus meridionalis (Rodentia, Sciuridae)]". -Hystrix. 28 (1). doi:10.4404/hystrix-28.1-12015.

- http://www4.uninsubria.it/on-line/home/articolo13602.html

South African moth in Portugal

A small, darkish brown moth from the southern hemisphere is now resident in central Portugal. There the species do not exhibit invasive behaviour, and so far has been only observed in very low numbers. The discovery is published in the open access journal *Nota Lepidopterologica* by an international research team, led by Martin Corley, CIBIO-InBIO, Portugal. In 2012, Jorge Rosete, one of the co-authors of the

study, spotted a female specimen that he could not identify near his house. When Martin took a look at it, he placed it in the concealer moth family (Oecophoridae), but was unable to recognise neither its species, nor its genus. It did not take long before a few more specimens were found, including males.

A fragment of DNA matched three other genetically identical unnamed specimens, originally collected from South Africa. Further collaboration with Alexander Lvovsky, Russian Academy of Sciences, allowed the assignation of the specimens to a species name: *Borkhausenia intumescens*, known from South Africa. However, it did not end there. Further research into museum collections showed that in fact this species had been previously described from Argentina as *Borkhausenia crimnodes*, and therefore should be named as such.

SOURCE: https://phys.org/news/2017-01-moth-europe-southern-hemisphere-species.html#jCp

Squeaky Troglodyte

Scientists from the Natural History Museum in Bulawayo, Zimbabwe have discovered the *Arthroleptis troglodytes*, a rare frog otherwise known as the "cave squeaker frog" in the Chimanimani Mountains of Zimbabwe 54 years after it was first discovered in the same location.

The elusive cave squeaker frog was first discovered in 1962 and was rediscovered for the second time on December 3, 2016 after a team of researchers mounted a thorough search for it. The frog was thought to have become extinct or critically endangered since it had not been found by anyone in over five decades, and is therefore considered one of the rarest amphibians still alive today.

SOURCES:

- http://www.itechpost.com/ articles/80932/20170206/cave-squeaker-frog-sighted-again-second-time-1962.htm
- http://www.inquisitr.com/3954074/ rare-cave-squeaker-frog-gets-sighted-for-first-time-in-over-five-decades/

Secret Seven

Four new frogs so tiny that they can sit on a thumbnail have been discovered in the forests of India. Among the smallest frogs in the world, they live on the forest floor and make insect-like calls at night. Three larger species were also found, bringing to seven the number of night frogs discovered in the Western Ghats.

The mountain range, which runs parallel to the western coast of India, is home to hundreds of threatened plants and animals. Scientists discovered the new species after several years of exploration in the forests of Kerala and Tamil Nadu. "These tiny frogs can sit comfortably on a coin or a thumbnail," said Sonali Garg of the University of Delhi, who was among the team that found the new creatures. "We were surprised to find that the miniature forms are in fact locally abundant and fairly common."

Another Compendium of Batrachia

Athirappilly night frog

"They were probably overlooked by researchers because of their extremely small size, secretive habitats and insect-like calls."

SOURCE: http://www.bbc.co.uk/news/science-environment-39042646

Send in the Clowns

Discovered in Bolivia: The new clown tree frog species *Dendropsophus arndti*, the new "flag ship" of the Senckenberg station "Chiquitos." . Credit: Senckenberg/Martin Jansen

An international team of scientists discovered two new species of clown tree frogs in the Amazon region. Until recently, these colourful amphibians had erroneously been considered part of another species. Now, DNA studies and an analysis of the calls of the examined populations revealed a much higher diversity within this group of frogs. Due to their small distribution areas, it is likely that the newly discovered species are threatened, but the determination of their protection status is currently still pending. In their study, published today in the scientific journal *PloS ONE*, the scientists from six countries clearly show that a complete species inventory is only possible by means of international cooperation.

In the past decades, more than 810,000 square kilometres of rainforest have been destroyed in the Amazon region, and every day, species from all animal phyla disappear from this area. "Our new study shows once again that we are not even close to knowing the actual species diversity of South American frogs and that even supposedly widespread species may be endangered," explains Marcel Caminer, the study's lead author from the Universidad Católica del Ecuador and he continues, "During expeditions to six Amazonian countries, we examined the two clown tree frog species *Dendropsophus leucophyllatus* and *Dendropsophus triangulum*, which were hitherto considered 'universal' species, in greater detail and were able to show that they do not constitute two, but at least five and perhaps as many as seven different species – two of which we were able to describe for the first time."

SOURCE: https://phys.org/news/2017-03-clown-tree-frognewly-threatened.html#jCp

Another Compendium of Batrachia

Honouring Dave

While there are already a number of species named after famous British broadcaster and naturalist Sir David Attenborough, including mammals, reptiles, invertebrates and plants, both extinct and extant, not until now has the host of the BBC Natural History's *Life* series been honoured with an amphibian. A new fleshbelly frog, recently discovered in the Peruvian Andes, is formally described as *Pristimantis attenboroughi*, while commonly it is to be referred to as the Attenborough's Rubber Frog. The new species is published in the open-access journal *ZooKeys*.

Scientists Dr. Edgar Lehr, Illinois Wesleyan University, and Dr. Rudolf von May, University of Michigan, spent two years (2012-2014) surveying montane forests in central Peru, in order to document the local amphibians and reptiles, and evaluate their conservation statuses. Their efforts have been rewarded with several new species of frogs and a new spectacled lizard.

SOURCE:
http://blog.pensoft.net/2017/03/07/new-frog-from-the-peruvian-andes-is-the-first-amphibian-named-after-sir-david-attenborough/

Freaky Fluorescent Froggie

Although this species has been known since the 18th Century, its secret has only just come to light (pun intentional). The South American polka dot tree frog (*Hypsiboas punctatus*) looks like a typical green-coloured frog under normal light. However, if you shine ultraviolet light on the amphibian, it lights up in a dazzling array of bright blues and greens. According to a new report in the *Proceedings of the National Academy of Sciences* journal, the South American frog is the first known amphibian capable of fluorescence, which is the ability to absorb light at short wavelengths and send light back out at longer wavelengths. In nature, fluorescence can be seen in various ocean creatures, including corals, sharks and sea turtles. It can also be seen in land animals like scorpions and parrots. Incidentally, fluorescence is different from bioluminescence, which is the capacity to generate light via chemical reactions.

SOURCE: http://www.nature.com/news/first-fluorescent-frog-found-1.21616

Another Compendium of Batrachia

Groovy Getaway Gecko

Many lizards can drop their tails when grabbed, but one group of geckos has gone to particularly extreme lengths to escape predation. Fish-scale geckos in the genus Geckolepis have large scales that tear away with ease, leaving them free to escape whilst

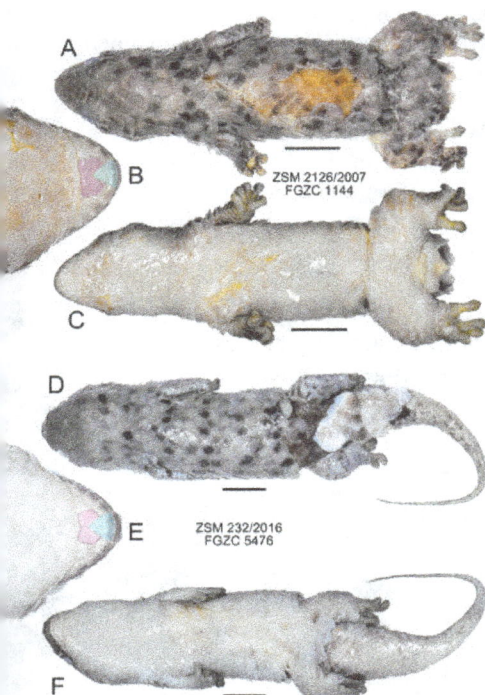

ZSM 2126/2007
FGZC 1144

ZSM 232/2016
FGZC 5476

the predator is left with a mouth full of scales. Scientists have now described a new species (*Geckolepis megalepis*) that is the master of this art, possessing the largest scales of any gecko.

SOURCE:
https://www.sciencedaily.com/releases/2017/02/170207092733.htm

Crabbe and Goyle

While not much is known about the animals living around coral reefs, ex-Marine turned researcher Harry Conley would often take to the island of Guam, western Pacific Ocean, and dig deep into the rubble to find fascinating critters as if by magic learnt at Hogwarts. Almost 20 years after his discoveries and his death, a secret is revealed on the pages of the open access journal *ZooKeys* -- a new species and genus of crab, *Harryplax severus*. Having dug as deep as 30 m into Guam's coral reef rubble, Harry Conley collected many specimens which stayed in his personal collection until the early 2000's when Dr. Gustav Paulay, currently affiliated with the University of Florida, handed the specimens to the second author of the present study, Dr. Peter Ng, National University of Singapore, which resulted in many discoveries and publications. Among the lot, however, were two unusual specimens which were not studied until much later. Only recently did Dr. Peter Ng and his colleague at the National University of Singapore and lead author of the paper, Dr. Jose Christopher E. Mendoza, discover that they represent not only a new species, but also a new genus.

SOURCE:
https://www.sciencedaily.com/releases/2017/01/170123125518.htm

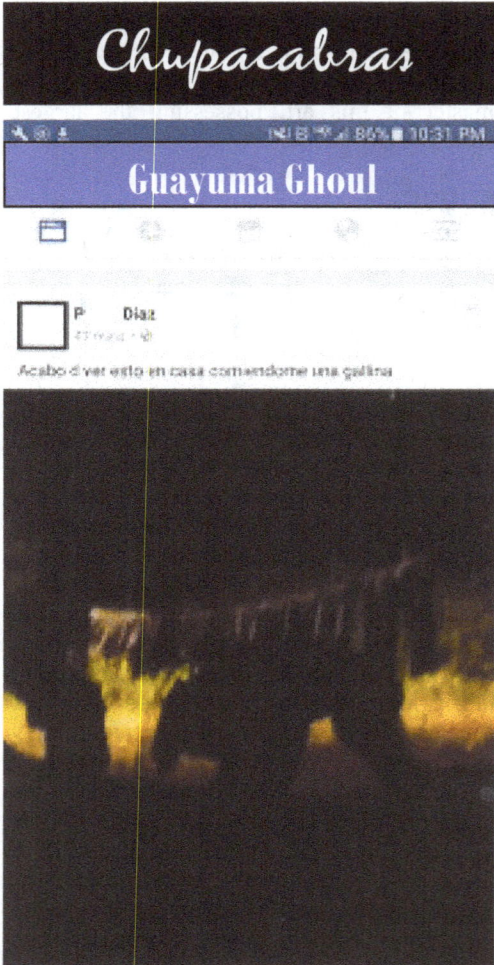

Chupacabras

Guayuma Ghoul

The above picture which apparently came from an unnamed Facebook page was sent to *Inexplicata* the blog of the *Journal of Hispanic UFOlogy* which is run by Scott

Corrales, who is a researcher that we have held in particularly high esteem for the last two decades.

They write that it "was taken on Monday December 25, 2016 at 10:31 p.m., in the Villodas sector of the town of Guayama, Puerto Rico, the same place where in 2013 were reported mysterious mutilations of animals."

The original post was captioned: "I just saw this at home eating a chicken". Bizarrely the post was taken down a few hours later, and Jose Perez, who wrote the article we have cited is understandably cautious.

However, he writes: "We want to clarify, that we do not know that it is the figure that is seen in the image, although we have to confess that it is extremely strange and curious we can not affirm that it is one of the anomalous entities that have been reported in countless occasions all around Puerto Rico and around the world.

We compare it to all breeds of dogs registered in the AKC and what you see in the picture does not match any of them. Neither can we identify with any other known animal."

SOURCE: http://inexplicata.blogspot.co.uk/2017/01/puerto-rico-strange-creature.html

Mexican Muddle

According to a report in *The Daily Mirror* in December:

"A horrifying demonic animal with long claws and three huge tusks has been spotted at a voodoo ground among the trees in Mexico.

The mystery animal was seen scurrying from a sinkhole near a graveyard - sparking fears it could be a blood-sucking vampire that is steeped in folklore across South America. The long-haired creature has baffled two vets that studied footage of it scuttling around in the forest in Yaxcopoil."

Later in the piece the author claims:

"But others also fear the mystery creature could have been conjured up through witchcraft after it was found at the site where numerous voodoo dolls and other things related to sorcery have been found.'

The video accompanying the story is so blurry that it is impossible to make out what this animal is. However it is obviously dead, and

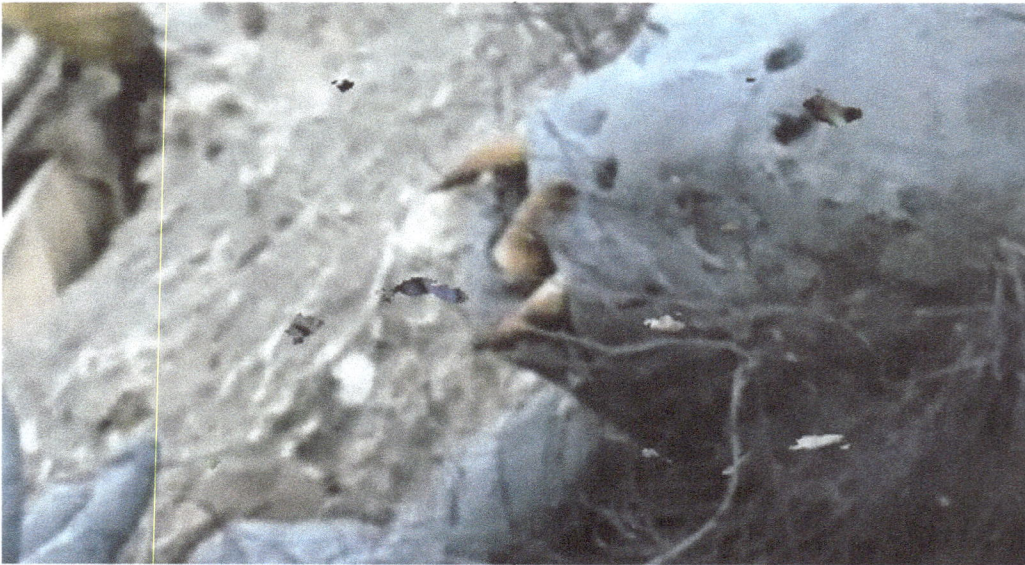

far from "scuttling" anywhere. The article also includes several grabs from the video, such as the one above, that I have truly no idea what it is. It also references the story of the unfortunate canid killed in the Ukraine in June that we covered in A&M57, together with new photos such as the one below, which—unfortunately—add nothing to the story.

Unfortunately, I can only deduce that this was the product of one of the slow news days just before Christmas.

SOURCE:
http://www.mirror.co.uk/news/world-news/demonic-creature-filmed-near-voodoo-9473585

Man Beasts (BHM)

Yalta Yeti

© YouTube / SDNikitin

In recent years, there have been more and more reported bigfoot-like creatures in Britain and Europe. Hell, your Editor and a CFZ investigation team saw something of the sort in Northumberland back in 2003. We have written about this elsewhere, and stated our opinion that the "thing" (as Ivan T Sanderson would undoubtedly have called it) could not be described or defined using zoological frames of reference.

We have not made ourselves particularly popular by saying that it is impossible for there to be an unknown species of higher primate living in the UK, but we will admit that the further one goes towards the almasty heartlands in the Caucasus, the more likely it could be.

Six years after the event, the *Daily Mail* reports on a sighting of a BHM in the Ukraine.

"A video a terrified hiker took of a Yeti-like beast has resurfaced online. Posted on the website Reddit the clip shows a 'Big Foot' hurtling through the woods in Ukraine. Filming was a petrified walker who didn't know what to make of what he was witnessing. You can hear him whimpering as he cowers in the forest. It is believed to have been filmed in Yalta on the Crimean peninsula that is contested by Russia and Ukraine."

The filmer captioned the YouTube clip: 'I don't know who or what it was.' But others argue the walker stumbled across the set of a film and the 'Big Foot' was nothing but an actor in costume. The video was first uploaded online six years ago and has since resurfaced on social media site Reddit in a discussion over the world's most convincing sightings of Bigfoot.

One wrote: 'Best part about this video is that the guy sounds genuinely scared.' Another said: 'Even if it is fake that's really good acting on that dude's part.'" Of course none of this is in any way conclusive, but it is interesting, and just like the *Daily Mail* we missed the story when it was first posted.

SOURCE:
http://www.dailymail.co.uk/news/article-4204446/Big-Foot-big-hoax-Terrified-hiker-s-footage-resurfaces.html

Snow Joke

A potentially interesting set of footprints ended up in the British press over the winter. Despite the fact that the story appeared in *The Sun*, it was treated with some degree of respect, and read: "The half-mile stretch of intriguing tracks were discovered in the snow in Sunnyslope, Washington state. Bigfoot researcher Paul Graves hopes the prints could be the long sought-after proof that Sasquatch is real. According to his observations, the tracks measure around two-foot long and around five inches wide at the heel, but they were strangely only two inches deep. The prints were around 4ft to 6ft apart, heel to heel, stretching in near perfect single file line from an orchard through a field and past an irrigation canal. Some have speculated that the prints may have been made by snowshoes".

Mr Graves commented: : "'There's no way that you could keep your snowshoe that in-line and for that far — all the way through that field, perfectly in-line.'"

SOURCE:
https://www.thesun.co.uk/news/2902083/do-mystery-two-foot-long-snow-prints-could-finally-prove-sasquatch-is-real/

The distance between each print varies from four to six feet... much longer than a human stride.

Mystery Cats

- **HERTFORDSHIRE**

Herts Police confirmed that between 2011 and 2016 there were 26 reports of big cats prowling in the wild. The majority (11) were of a panther, with one informant, in 2011, saying that after seeing a sheep being eaten by a panther, they "ran all the way home and our dog was going crazy, barking and yelping". Of the 26 reported sightings, nine were classified as unknown species, but shared similar characteristics with callers telling the police the big cats were "very large" or the same size as a Labrador, brown or gingery in colour – and in one incident, just 20ft from the informant.

SOURCE: http://www.whtimes.co.uk/news/big_cats_prowling_welwyn_hatfield_freedom_of_information_request_reveals_1_4793383

- **SCOTTISH HIGHLANDS**

A large "cat" with a taste for mutton is believed to be roaming the most remote parts of northern Scotland, stripping sheep from their skin (below) and leaving no trace but

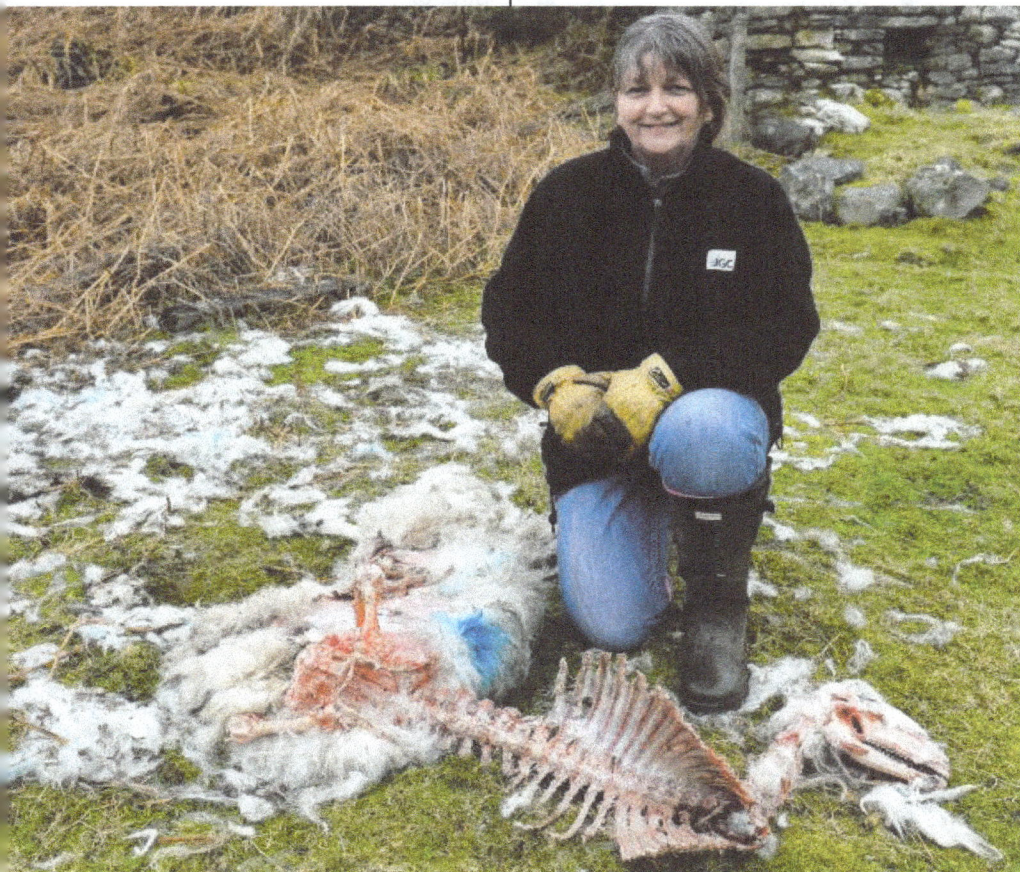

bones and wool. The mysterious beast's latest catch is a hefty and healthy ewe of about 50kg. Her wool was neatly peeled off her skin before it was eaten. The carcass was found less than 100 yards from a croft in Swordly at the weekend.

SOURCE: http://www.independent.co.uk/ news/uk/home-news/beast-cat-scottish-highlands-bettyhill-swordly-sheep-stripped-eaten-mystery-a7616711.html

• **YORKSHIRE**

A mystery cat (below) carrying unique leopard spot markings has been spotted in the woods near to Robin Hood's Bay. The animal is estimated to be at least three feet long in length, prompting the assertion that it may be a hybrid wildcat. Some have suggested that the cat may in fact be a Bengal cat, which carries leopard spot markings. The footage was captured on a wildlife camera by Alan Adams from Skerry Hall Farm Bed and Breakfast, based in the Bay.

SOURCE:
http://www.whitbygazette.co.uk/news/video-unique-looking-cat-spotted-in-the-bay-1-8411795

• **DERBYSHIRE**

36-year-old Matt Chambers said he was driving along the A38 when he spotted a large black cat between Egginton and Stretton.

SOURCE:
http://www.derbytelegraph.co.uk/black-panther-sighting-in-burton-every-big-cat-reported-to-derbyshire-police-between-2007-and-2017/story-30218094-detail/

Aquatic

D:019.1M
H:241

Britain's rarest fish spotted

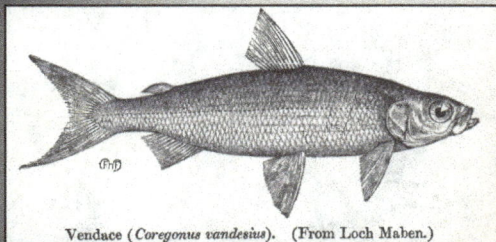
Vendace (*Coregonus vandesius*). (From Loch Maben.)

THE UK's rarest freshwater fish - a relic of the ice age - has been caught on film for the first time during routine survey work in the Lake District.

The fleeting shot of the elusive vendace, an international conservation priority, was captured using a remote-controlled yellow submarine on the bottom of Derwent Water. The film was taken by the Environment Agency and the Centre for Ecology & Hydrology, who are working to assess how much sediment is building up in the body of water.

SOURCE:
http://www.thewestmorlandgazette.co.uk/news/15053728.UK_s_rarest_freshwater_fis h_caught_on_camera_in_Lake_District/

Most authorities now consider *Coregonus vandesius* to be a subjective synonym of *Coregonus albula*, which is a more widespread North European freshwater whitefish species. Both taxa are also known by the common name vendace. The status however remains controversial, and FishBase still lists *C. vandesius* as a separate species, reflecting the recent treatment of the European freshwater fish fauna by Kottelat & Freyhof (2007). Another synonym of *British C. vandesius* is *C. gracilior*.

The vendace has only ever been known as a native species at four sites in Britain: Bassenthwaite Lake and Derwent Water in the English Lake District, and the Castle Loch and Mill Loch in Lochmaben, Scotland.. The Castle Loch population disappeared in the early part of the 20th Century, and the Mill Loch population disappeared in the 1990s. The fish had not been recorded at Bassenthwaite Lake since 2001, but was rediscovered in 2014.

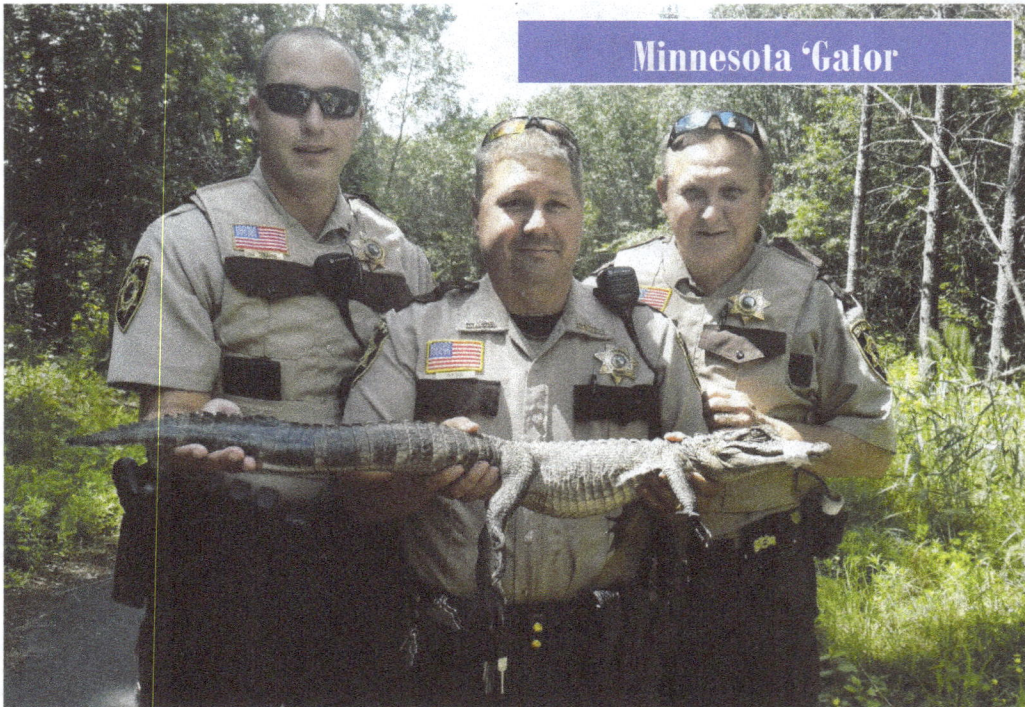

Minnesota 'Gator

The American alligator is the widest distributed crocodilian in North America, (the others are the American crocodile, plus feral populations of the spectacled caiman in Florida). American alligators are found in the wild in the south-eastern United States, from the Great Dismal Swamp in Virginia and North Carolina, south to Everglades National Park in Florida, and west to the southern tip of Texas. They are found in North Carolina,

South Carolina, Georgia, Florida, Louisiana, Alabama, Mississippi, Arkansas, Oklahoma, and Texas. Some of these locations appear to be relatively recent introductions, with often small but reproductive populations.

For those of you not versed in American geography (as Morissey said "America is not the world") the photoshopped sketch map below shows in green the approximate range of this species.

The state coloured in red is Minnesota, which is obviously far outside the natural range of the species. But, last summer there was a string of alligator reports there.

Law enforcement in Cass and Crow Wing counties have taken reports of three separate alligator sightings in the past two weeks — and

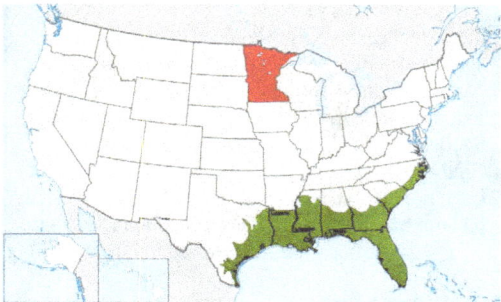

managed to catch two of them. It's a striking uptick in alligator sightings, according to Cass County Sheriff Tom Burch, considering he passed the bulk of his long career with no sightings at all. "We deal with bears at certain times of the year," he said. "Occasionally we get cougar sightings. This is the first time I recall ever dealing with an alligator."

Early Tuesday afternoon, Cass County deputies landed a 30-inch alligator on Hardy Lake with a fishing net and the help of a Minnesota Department of Natural Resources conservation officer. The young alligator was first spotted Monday on the north end of the lake, sunning itself on the beach. Burch said if the alligator was much bigger, his deputies wouldn't have known what to do. "I might be OK if it's a couple feet long," he said. "But if it's 6 or 8 feet, I'm going the other way."

Before that, a report came in Wednesday of another alligator in Sylvan Township.

This little run of alligator sightings started last week on a bike trail south of Brainerd. A cyclist spotted a small alligator crossing the trail and called 911. The Crow Wing County Sheriff's Office sent out three deputies to investigate. "They were actually pretty excited about it," said Lt. Joe Meyer. "It's not every day you come across an exotic animal like this.'"

All of these reports are almost certainly of released pets, but if the planet continues to warm like it is doing, the idea of alligators increasing their range inexorably northwards is not that unlikely a scenario.

SOURCE:
http://www.mprnews.org/story/2016/07/19/ brainerd-lakes-authorities-stay-away-from-alligators

Morgawr Returns?

The story broke in mid-February when a story in *The Metro* proclaimed:

"A 'serpent' carcass has stunned walkers

who found it washed up on a beach. The eight foot skeleton has mostly rotted away, but the shape of its body, which bulges out towards the back, is still visible. People think it could be the legendary Cornish monster known as Morgawr, said to live off the coast of Falmouth. Drawings of the serpent, whose name means sea giant, indicate it could have a similar shape with its body tapering down to a narrow neck, rather than the more stocky bodies of dolphins and whales. Chris Crane, 61, was walking with wife Amanda, 59, when they saw a large black figure in the distance on Charlestown beach, close to St Austell."

Blah blah blah.

It is, of course, the decomposed and very manky carcass of a pilot whale, just like the

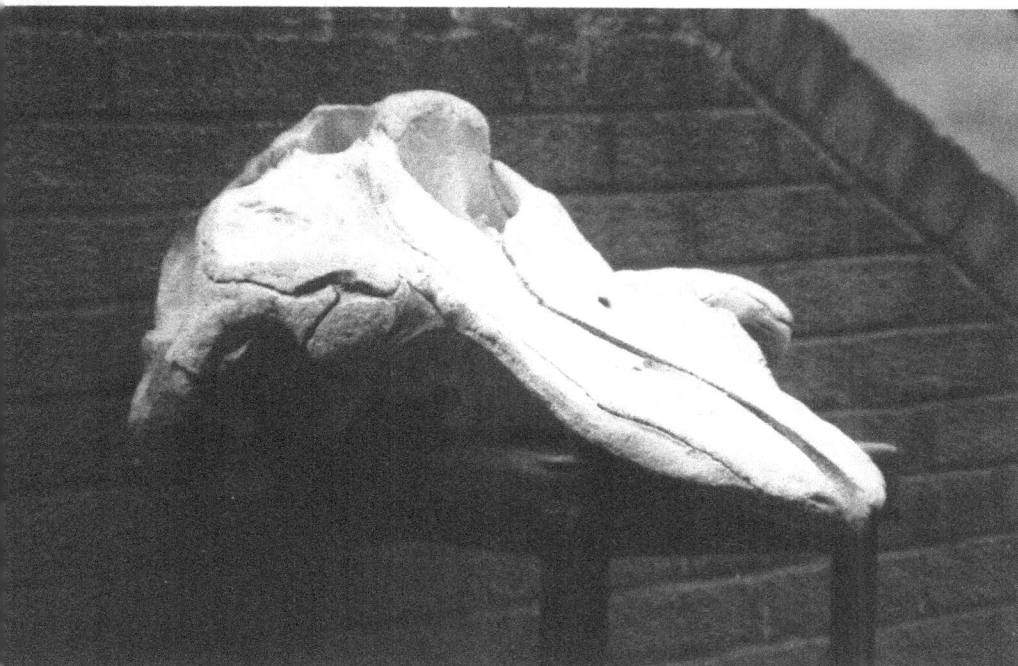

one which was washed up on Durgan Beach, Cornwall at Christmas 1975, (above) and was touted far and wide as being the carcass of morgawr. At the time, 'Doc' Shiels said that it was now't but a whale, and when we recovered the skull years later we confirmed that.

Darren Naish confirmed in 2008 that it was the skull of a long finned pilot whale, and the latest "Morgawr" pictures suggest that this latest carcass is more of the same.

Truly, there is nothing new under the sun!

SOURCE:

http://metro.co.uk/2017/02/23/skeleton-of-eight-foot-serpent-washes-up-in-cornwall-6467504/

HERALD OF EARTHQUAKES

An interesting piece of 21st Century folk belief emerged earlier this year. Just a few days before the "polar bear" carcass washed up in the Philippines (see Newsfile Xtra #2)

ABS/CBN News ran the following story: "The earthquake that jolted Surigao del Norte on Friday night jogged the memory of some people on social media, who were saying an incident two days prior could've served as a sign for the disaster that was about to strike.

On Wednesday, a 10-foot-long dead oarfish was caught by fishermen off the coast of Agusan del Norte.'

To some, deep-sea creatures such as oarfish that end up in shallow water are a good predictor of earthquakes. How true that is remains very much a debate? "It's theoretically possible because when an earthquake occurs there can be a build-up of pressure in the rocks which can lead to electrostatic charges that cause electrically charged ions to be released into the water," Rachel Grant, a lecturer in animal biology, said in a report posted on the Independent news website in October 2013.

In an undated Japan Times article, seismologist Kiyoshi Wadatsumi said that "deep-sea fish living near the sea bottom are more sensitive to the movements of active faults than those near the surface of the sea."

SOURCE: http://news.abs-cbn.com/ news/02/11/17/dead-oarfish-found-in-mindanao-sparks-debate-can-animals-predict-earthquakes

Newsfile Xtra

The Scottish "Polar Bear"

When I was a boy, I used to read a magazine called *Look & Learn*. It included a comic strip, the only details of which that I can remember, being that as a sub plot, it had polar bear cubs drifting down from the Arctic to one of the more remote Scottish islands.

I always thought that this was massively cool, but, as far as I am aware it has no basis whatsoever in truth. *The Daily Star* of the 15th of November claims that dead polar bears have washed up on the beaches of Colonsay in the Scottish Inner Hebrides. Have my boyish dreams been answered? Probably not. But, I think that this is an interesting case from a cryptozoological point of view.

In October 1924 a mysterious carcass was washed up on Margate Beach, South Africa and stayed there for 10 days. It was described as being covered in snowy white fur, having a trunk like that of an elephant, the tail of a lobster and being completely empty of blood.

In the intervening years all sorts of people have made all sorts of peculiar suggestions as to what this creature might have been. My favourite was that it was a

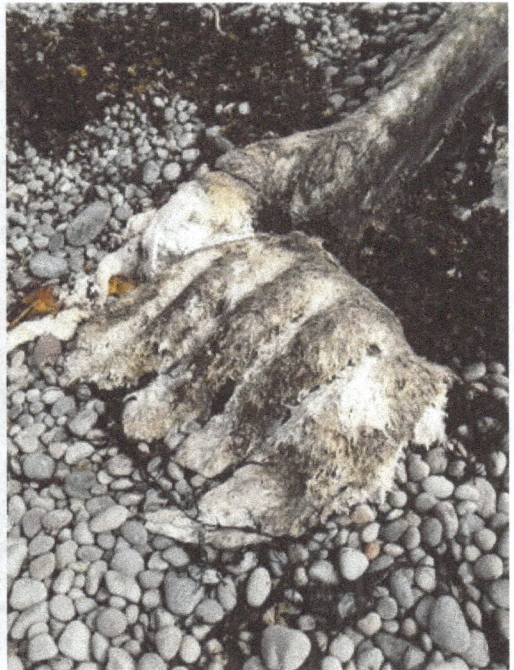

shown in the first photograph, would not still be so clearly defined as it is, considering the rough handling to which it must have been subjected by the sea.

"At one end there is a round lump about two feet in diameter, which might be taken for a head, but there are neither eyes, mouth, ears, nor anything else visible. There are, moreover, no limbs, flappers, tentacles,

The extraordinary creature washed up at Margate, Natal. It was about fifteen feet long, but apparently possessed neither head, limbs, tentacles, nor fins.

tail, or any other features which would help to identify it. The measurements are as follows: About fifteen feet long, six feet broad, and two feet thick. The carcass, at the moment of writing, has already begun to decompose, and there can be little doubt that it is composed of flesh of some sort. Probably, if it is allowed to rot and the skeleton becomes visible, it will be possible to identify it by this means.

"Some time ago a monster of unknown species was washed up on the east coast of South America after a submarine earthquake, and it was concluded

STRANGE SEA FIGHT was observed off the coast of Natal, South Africa, by a resident of Margate. The sea was quite calm, and he saw what he took to be two whales fighting some sea monster. He got his glasses and was surprised to see an animal which resembled a polar bear, but, in size, was equal to an elephant. This object he observed to rear out of the water fully 20 feet and strike repeatedly at the two whales, but with, seemingly, no effect. The fight went on for fully three hours, gradually nearing the shore, but it grew too dark to make further observations. Next morning the man found the monster high and dry on the beach. He walked right round the creature, but could find no head! Where the

The newspaper cutting describing the sea-fight.

just off the coast, that the creature had been thrown up from some hole in the ocean bed. A little while before this mysterious animal put in its appearance on the Natal coast an earthquake was registered at Durban, and was calculated to have occurred beneath the ocean near Margate. It is thought that the monster may have been dislodged from the ocean-bed in the same way as the South American specimen. The only other suggestion put forward is that the monster has been carried down from the Arctic regions!

"It may be of interest to know that quite a number of small pieces of pumice-stone have lately been picked up along this part of the coast. Possibly the earthquake may account for both the pumice-stone and the monster?"

Another reader sends us the attached cutting, from a local newspaper, which is interesting as describing the fight referred to by Mr. Jones. The newspaper says that "the monster's backbone was very prominent and the whole body covered with snow-white hair."

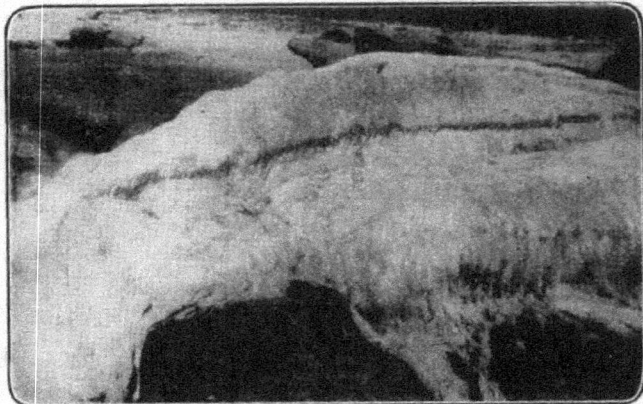

A closer view of part of the monster, showing the slimy-looking white hair.

hitherto unknown species of marine elephant, although I would admit I didn't take it very seriously, and I winced when the creature became widely known as 'Trunko'..

On 6 September 2010 Karl Shuker announced that a hitherto-unknown photograph of Trunko had been discovered by German cryptozoologist Markus Hemmler on the website of the Margate Business Association, and Shuker recognised from this photo that Trunko had been nothing more than a globster, i.e. a massive, tough skin-sac of blubber containing collagen that is sometimes left behind when a whale dies and its skull and skeleton have separated from the skin and sunk to the sea bottom. The photo had been snapped by Johannesburg photographer A. C. Jones, who had visited Trunko's remains while they were beached.

Three days later, Shuker revealed that he and Hemmler had independently discovered two more photos of Trunko by Jones that had been published in the August 1925 issue of *Wide World* magazine. These close-up photos showed a

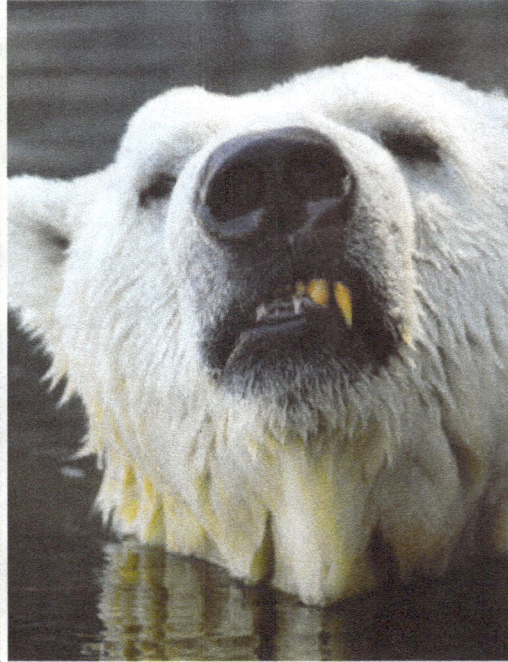

classic globster, confirming Shuker's identification of Trunko, and clearly revealed its white 'fur' to be exposed connective tissue fibres. It was the sight of two whales some distance out to sea tossing this globsterised mass into the air, a common practice, that had fooled observers on Margate Beach into assuming that it was alive. Perhaps the most surprising aspect of this revelation is that two photos of the Trunko carcass had been published in a mainstream magazine in 1925, yet had somehow been entirely overlooked afterwards by the zoological and cryptozoological community for the next 85 years.

More recently, there have been fevered arguments between Karl Shuker and Max Hawthorne as to whether this carcass was a small bona fide globster or the decaying body of a white or whiteish bull right whale but everybody seems to concur that the 1924 carcass was cetacean in origin.

Now we have so–called polar bear carcasses washing up in Scottish beaches. I think that it is beyond contention that they are actually similar Oregon to the 1924 carcass.

This is where I am frustrated that the CFZ does not have the money to do what I want it to do. I would love to be able to send somebody up to the Scottish Islands and take samples, so that we could put this story, and quite possibly the 1924 story, to bed once and for all.

Newsfile Xtra II

But there is more....

The previous Newsfile Xtra originally appeared in December's CFZ Members Newsletter, but at the end of February a startlingly similar story arrived in our inbox, this time from The Philippines.

"The carcass of a monster-size sea creature has washed ashore in the Philippines, and the mystery today is not only focused on what it is, but also what caused the death of this entity covered in white hair. This is the second carcass to wash ashore in the Philippines recently, and both carcasses look like sea monsters to locals.

The two very different looking monster-like carcasses washed ashore at separate locations and days apart. To locals, it appears that something is killing the creatures that dwell in the deepest parts of the ocean, which are places humans seldom see.

The Philippino "Polar Bear"

According to *The Sun* a "massive hairy

blob-like monster" washed up on a Philippine beach, and the pictures of this mass are splashed across the social media sites as people stopped by to take selfies with this creature. The creature appeared to be bleeding, so it was apparently once a living thing. The deep sea monster-like entity washed up in the Cagdainao, Dinagat Islands.

Because of the massive amount of white hair-like substance on the carcass, some thought it was a massive polar bear."

Once again the 'polar bear' identification is used, even though these islands are even further from the true habitat of polar bears than the north of Scotland, where the other specimens were found.

Once again, our verdict is that this is another dead whale, or part thereof exhibiting the results of decomposing collagen.

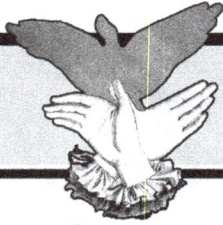

Cryptozoology

'Extinct' birds to fly from new habitat

HYDERABAD: Jerdon's Courser (*Rhinoptilus bitorquatus*) pictured above, and the great Indian bustard (*Ardeotis nigriceps*), the critically endangered birds of Andhra Pradesh, got a new 'lease of life' with priority being given to these avian species in the National Wildlife Action Plan 2017-31. The birds were included in a five-year species recovery plan, which ended last year, but now an alternative habitat for these birds will be identified and developed in the next four years to save these birds from extinction.

Jerdon's courser is endemic to Andhra Pradesh and thrives only in Lankamalleswara Wildlife Sanctuary in Kadapa district. There is only a handful of Jerdon's coursers believed to be alive; the bird was last seen a decade ago. This species became `extinct' and was `rediscovered' twice in the last 180 years. It went extinct in 1846 and 1900, but sighted in 1986.

The great Indian bustard is also critically endangered and lives in six states in India, including Andhra Pradesh. The number of birds in its exclusive sanctuary, Rollapadu in Kurnool district, has seen a sharp dip, according to researcher Mohammed Ghouse of Osmania College in Kurnool.

Though the great Indian bustard (pictured below by L R Burdak) was covered under the previous national wildlife action plan that

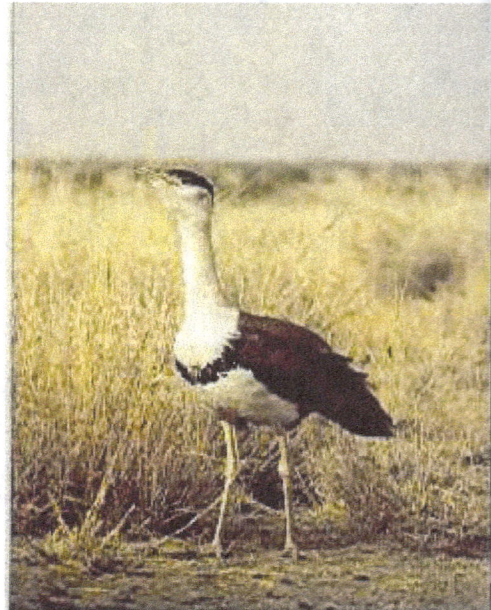

ended in 2016, the Centre could not carry out studies to evaluate the ecosystem services of this fragile bird species. Mohammed Ghouse said, "Less than 300 birds exist in India. Thanks to agricultural activity, the habitat is disturbed. This resulted in decline of the Great Indian Bustard population." This species is hunted, which has also led to its decline, as well as unchecked encroachment of the bird's natural habitat. Researchers also attribute the decline in numbers to increased human activity; the shy birds are known to leave their nests when disturbed.

Source: http://timesofindia.indiatimes.com/city/hyderabad/extinct-birds-to-fly-from-new-habitat/articleshow/56770355.cms

Conservation of ivory-billed woodpecker's habitat urged

According to a new study published in Heliyon, the ivory-billed woodpecker's (*Campephilus principalis*) habitat should be protected despite the lack of definitive evidence of its existence. At the moment, bird conservation efforts rely on indisputable photographic evidence, which according to the new study could take many years to obtain, by which time it may be too late.

Threatened by habitat destruction and other factors, this species has been declared extinct only to be rediscovered several times. In the absence of indisputable evidence, the discourse on the bird's existence has been dominated by opinion. After ten sightings during an eight-year search, Dr. Michael Collins of the Naval Research Laboratory in the US believes the ivory-billed woodpecker is alive -- but the bird needs our conservation efforts now, regardless of the proof, if it is to survive.

They live in vast swamp forests - Florida and Louisiana, in particular - which are difficult and dangerous to access. Apart from the threats of dangerous animals, there is a risk of being accidentally shot in areas that are heavily hunted, and most bird watchers will never visit the bird's habitat. The woodpeckers are highly elusive and wary of human contact, hiding away and keeping quiet at the first sign of threat. Using these behavioural and habitat factors, Dr. Collins has been able to approximately quantify the elusiveness of this bird, concluding that it would take significantly longer to photograph the ivory-billed woodpecker than similarly rare North American birds.

"There is no logical reason to require a particular form of evidence," said Dr. Collins. "When faced with an exceptional case, scientists often develop alternative approaches and make progress using different types of data."

In the paper, Dr. Collins presents three videos -- one of more than 20 minutes -- that show birds he believes to be the ivory-billed woodpecker as they have many characteristics consistent with the bird but no other species living north of Mexico. The bird's remarkable swooping flights, rapid wingbeats, and an audible double-knock are captured on film, consistent with reports from the 1940s and earlier. He observed the ivory-billed woodpecker ten times in 1500 hours of searching between November 2005 and June 2013.

"Having observed these birds is one of the two most deeply meaningful experiences of my life. When I was 11 years old, I stood in my front yard in Tampa, Florida, and watched Apollo 11 blasting off into space on the way to the first manned landing on the Moon. I feel very privileged to have been a direct eyewitness to a symbol of the vanishing wilderness of our world as well as one of the great achievements of humankind. My hope is that we will continue making progress and doing great things while at the same time preserving our natural world."

Source: https://www.sciencedaily.com/releases/2017/01/170124111423.htm

New immigrants in Nepal

In the mountainous region of Nepal, a new species of birds has been discovered, taking the overall number of avian species in the Himalayan country to 866.

A rufous-tailed rock thrush (*Monticola saxitilis*), seen as an autumn channel migrant bird species in Pakistan and India, photographed nearby Shey monastery within the Shey Phoksundo National Park at voyage last year. Investigators from an NGO Friends of Nature (FoN) Nepal covered the bird while reviewing Himalayan wolf, wild yak and snow leopard. The proper identification of the bird informed by the squad was completed by Carol Inskipp and Hem Sagar Baral, who are the bird experts.

The Branch of National Parks and Wildlife Conservation (DNPWC) and Bird Conservation Nepal sanctioned the presence of a new bird species in the nation, said Naresh Kusi, from FoN.

Source: https://www.newdelhitimes.com/new-immigrants-in-nepal123/

Guest from Mongolia flies to Gangapur after 20 years

An osprey (*Pandion haliaetus*) was sighted in Gangapur dam recently after nearly 20 years. This bird is known to come all the way from Mongolia and at times flies to South Africa. The osprey is also known as the fish eagle and is a 60cm in length and 180cm in width (across the wings).

"The bird comes from North Europe. It

travels from Mongolia and goes as far as South Africa. We had spotted it last around 20 years ago. It is a beautiful bird weighing between 1.4kg and 2.5kg. We saw the male osprey, which is slender and has a faint collar in the neck. The female bird has a dark band around the neck," honorary wildlife warden Bishwaroop Raha said.

He said the bird must have migrated there as the water of Gangapur is deeper than that of Nandur Madhyameshwar, where it is shallow, and it dives deep in search of fish. "It dives from 50feet height for fish. It is not a rare bird, but its sighting is very rare in Gangapur dam. All over the world, there are around 2.50 lakh ospreys," Raha added.

Source: http://timesofindia.indiatimes.com/city/nashik/guest-from-mongolia-flies-to-gangapur-after-20-years/articleshow/57023087.cms

Rare owlet believed extinct for 113 yrs thriving in Dang

The farmlands of Dang are home to as many as 82 forest owlets (*Athene blewitti*), pictured here

by Nimtan, the critically endangered species whose numbers are estimated to be just 200 to 400 individuals globally. What's even more encouraging is the numbers of this bird, which is endemic to central Indian forests, may increase in Gujarat.

In September 2014, Jenis Patel and his team of volunteers from Voluntary Nature Conservancy (formerly known as Vidyanagar Nature Club) had first recorded the presence of forest owlet in Purna Wildlife Sanctuary, breaking the long-standing belief that the bird has vanished from the dense forest of Gujarat. For nearly 113 years, the forest owlet was thought to be extinct until researchers rediscovered it in 1997 in Toranmal Reserve Forest.　Later their presence was thought restricted to Maharashtra and Madhya Pradesh.

"We have concluded that forest owlet prefer a habitat near human settlement in Gujarat. It is more like farmers' owlet instead of forest owlet," said Jenish.

Source: http://timesofindia.indiatimes.com/city/vadodara/rare-owlet-believed-extinct-for-113-yrs-thriving-in-dang/articleshow/56382553.cms

Conservation and Welfare

Foie Gras Production Has Been Banned in the 'Capital of Europe'

As the home to many European Union institutions, Brussels is setting a precedent for the demise of the hideously cruel industry of foie gras production by announcing a ban in the Brussels-Capital Region.　Brussels' Minister for Animal Welfare said, "[Foie gras production] is truly a kind of torture imposed on ducks, and we can hardly tolerate it".

It's a huge step forward for Brussels – known unofficially as the "Capital of Europe" – to take a stand against this abusive industry, and PETA has written to Belgium's Minister for Agriculture to urge him now to ban production across the country. Belgium is one of five European countries – alongside Bulgaria, France, Hungary, and Spain – still producing foie gras. Part of PETA's letter reads as follows:

Foie gras production is illegal in the UK and more than a dozen other European countries – including Germany, Luxembourg, and the Netherlands – and more and more countries around the world are outlawing the cruel practice of force-feeding birds. It's high time that Belgium followed suit.

Source: http://www.peta.org.uk/blog/foie-gras-production-banned-capital-europe/?utm_campaign=VEG%20foie%20gras%20victory&utm_source=ENews%20Email&utm_medium=Promo

Is Australia the birthplace of birds' nests?

The most common birds' nests found today had their birthplace in Australia, and these nests may be key to many of our birds' success, according to new research.

The research, published in *Proceedings of the Royal Society B*, looked at the nests of passerines – which include lyrebirds, fairy-wrens and magpies – and found the ubiquitous 'open' cup nest evolved in Australia multiple times more than 40 million years ago.

The birds resulting from these open cup nesting lineages gave rise to many of the worlds' birds today, and this study suggests that the open nests were perhaps a key to their success – measured both in terms of how many species they evolved into, and how far they have spread around the world.

"Among the passerine birds – which make up 60 per cent of the worlds birds – most species today build open cup-shaped nests, and only a minority build more elaborate roofed structures. The study shows that open cup nests evolved multiple times independently during early passerine evolution on the Australian continent, eventually becoming the most common nest type across the world today," said study co-author Professor Simon Griffith from the Department of Biological Sciences.

"Australia is host to the ancestors of today's common birds around the world, and the open cup nest that originated here is one of the innovations that perhaps has made them so successful," Professor Griffith added. "Until now we had assumed that more complex fully roofed nests had evolved from those without roofs.

This study demonstrates that in fact it was the opposite, in that these simple nests evolved several times independently, and the bird families that made this switch to simple nests are some of the most species-rich bird families today, such as the Australian honeyeaters."

He concluded, "This research really underlines the importance of Australia as the source of much of the worlds' avian diversity – Australia was the birthplace for which many key features of birds started out and still holds representatives of many of the ancient families."

Source: http://www.rarebirdalert.co.uk/v2/Content/Is-Australia-the-birthplace-of-birds-nests.aspx?s_id=97764830

Threatened seabird successfully breeds using artificial nests for first time

The Japanese murrelet (*Synthliboramphus wumizusume*) pictured here by Alistair Rae is a small seabird with an equally small range; it can be found only in warm current waters close to Japan. The birds' breeding range is even smaller, being concentrated mainly on the ground of rock reefs or isolated islands from Kanto region and to the west, where they make their nests in the crack of rocks. Two days after hatching, the chick abandons the nest for the sea and is fed by its parents off the coast.

Although little is known about the species' ecology because the bird lives at sea almost all year round, it is known that the population is decreasing rapidly, and as a result of these declines, the Japanese murrelet has been designated as a vulnerable species both by the Ministry of Environment of Japan, and by BirdLife International, who have assessed the species' conservation status on behalf of the IUCN Red List.

The main causes of the decreasing numbers have been shown to be the deterioration of the breeding sites, disturbance by people, and predation of the adult birds and chicks and eggs by rats and crows. The latter threat is a result of human activity in the birds' breeding range. In response to these declines, Wild Bird Society of Japan (BirdLife Partner) has been working for conservation of the bird, with one of their main aims being to designate a protected area of breeding sites and improve the environment of them.

Since 2010, in an attempt to improve conditions at the breeding sites, WBSJ has been developing prototypes of artificial U-shape concrete block nests, but as the Japanese murrelet had never been known to use artificial nests before, there was no prior knowledge of how to construct an artificial breeding site that would meet the bird's requirements, so WBSJ had to experiment with different designs. After six years of trial and error, finally chicks of the Japanese murrelet fledged from three artificial nests in 2016.

Hironobu Tajiri, Ph.D, Manager, Preservation Project, WBSJ said, "If we improve more artificial nests and standardize them, we could install them on islets where the bird population is decreasing or the bird can't be seen. It could be possible to have breeding birds there again".

Source: http://www.rarebirdalert.co.uk/v2/ Content/Birdlife-Japanese-Murrelet-successfully-breeds-using-artificial-nests-for-first-time.aspx?s_id=97764830

British military bases on Cyprus are Europe's deadliest place for birds

Anti-poaching organisation, the Committee Against Bird Slaughter (CABS), along with the Foundation Pro Biodiversity (SPA), has today revealed the massive scale of bird trapping in the UK-governed Sovereign Base Area (SBA) in Cyprus. More than ten kilometres of illegal nets were detected in just two nights and 274 nets were removed.

This once again reveals that the Sovereign Base Areas are probably the deadliest place for birds in the whole of Europe. More importantly, it also demonstrates the fact that the responsible authorities (UK-governed SBA police) are unwilling or even unable to handle the issue, despite assurances made two months ago.

Alex Heyd, General Director of the Committee Against Bird Slaughter said: "The situation in the SBA is now worse than ever. Every time we monitor trapping in the Base the same well known trapping sites are active, with tape lures – recorded bird calls to attract others to the site - that are detectable from miles away. The same number of nets, if not more, are always found. If you consider that trapping sites are often only 50 yards away from each other and nets can be seen from roads and even the highway, the only possible conclusion is that the enforcement efforts of the SBA police to tackle this issue is unsatisfactory, to say the least."

Even though the SBA Police has an 11-member unit dedicated to combat trapping crime, they have caught and fined only 62 poachers for illegal bird trapping in the last two years, on average one every six days. Compared to CABS/SPA findings, estimating some 150 trapping sites being active every day during the trapping seasons (autumn and winter), one can easily make the assumption that the SBA Police only scratches the tip of the iceberg.

"CABS/SPA is more than willing to logistically and strategically assist the SBA Administration in tackling bird trapping, but every year we are left wondering whether the SBA Police is really willing to tackle bird crime in their jurisdiction" concludes Mr Heyd.

Source: http://www.rarebirdalert.co.uk/v2/content/British-military-bases-on-Cyprus-are-Europes-deadliest-place-for-birds.aspx?s_id=97764830

Mapping movements of alien bird species

The global map of alien bird species has been produced for the first time by a UCL-led team of researchers. It shows that human activities are the main determinants of how many alien bird species live in an area but that alien species are most successful in areas already rich with native bird species.

"One of the main ways humans are altering the world is by moving species to new areas where they do not normally occur. Our work shows why humans have been moving these 'alien' bird species around for the last 500 years -- primarily through colonialism and the increasingly popular cage bird trade ¬- and why some areas end up with more species than others," explained supervising author, Professor Tim Blackburn (UCL Genetics, Evolution & Environment and ZSL).

For the study, published in *PLOS Biology*, the researchers collected and analysed data on the movement of almost 1,000 alien bird species between 1500 and 2000 AD. This was used to create a new open access database which was then analysed for patterns in the context of historical events and natural environmental variation.

More than half of all known bird introductions were found to occur after 1950, likely driven by the cage bird trade, and the researchers say this trend is expected to continue.

Source: http://www.rarebirdalert.co.uk/v2/Content/Mapping-movements-of-alien-bird-species.aspx?s_id=97764830

Killer powerlines taking their toll on Africa's birds

Africa is powering up rapidly. Governments' urge to develop economically and attract investments is immense. This push also involves countries along the Red Sea/Rift Valley Flyway Region including Sudan, where millions of migratory birds could be dying due to electrocution and collisions with electricity cables in an expanding power sector.

A preliminary report from recent surveys in Sudan has revealed that eight bird species have been mostly affected by the powerlines. These include Black Kite, Lesser Kestrel, Common Kestrel, Yellow-billed Kite, Abdim's Stork, Grayish Owl, White-backed Vulture and Pied Crow. The surveys were conducted in Al Gazeira, Al Gadarif and Kassala States in June/July and December 2015, as well as January 2016. The powerlines surveyed in Kassala State proportionately recorded the highest number of birds killed. These surveys were commissioned through the Migratory Soaring Birds project of BirdLife International funded by GEF/UNDP.

Source: http://www.rarebirdalert.co.uk/v2/ Content/Bird-killing-powerlines-in-Sudan.aspx?s_id=97764830

Fewer gulls, more mink and nutrients implicated in pochard decline

The number of common pochard (*Aythya ferina*) pictured above by Richard Crossley, migrating to the UK for winter has decreased by 60% since the 1980s, despite the number that breed here doubling over the same period. This decrease reflects the widespread declines in breeding numbers recorded elsewhere across Europe, of which little has been understood until now.

Now reasons for the recent decline of common pochard are investigated for the very first time in an issue of *Wildfowl*, the international scientific journal of the Wildfowl & Wetlands Trust (WWT). The

paper by 29 researchers from across Europe describes the changes that have most likely affected the pochards:

Fewer gulls - colonies of nesting black-headed gulls provide perfect cover from predators for other nesting birds, and their widespread disappearance is known to have affected pochards' breeding success. More nutrients - nutrients washed off farmland prompt explosions of plants and algae in the wetlands and waterways, preventing pochards and other birds diving for food.

More predators - the escaped American mink was introduced to Europe for fur and, along with fellow invasive aliens raccoons and raccoon dogs, have become major wetland predators, killing pochards and other creatures in great numbers. Wildfowl Editor and WWT Research Fellow Dr Eileen Rees said: "Shedding light on the problems such as those facing the Common Pochard is why Wildfowl is so important. Waterbirds live complex lives and their wetland habitats are entwined with our own need for food and water. Waterbird populations are often spread across nations and continents, and they're highly mobile, breeding, staging and wintering in different places. Wildfowl has long been acknowledged as an international scientific journal dedicated to wetlands and the birds that inhabit them. We're able to report the concerted efforts of colleagues from institutions across the globe to help our understanding of these magnificent birds. I'm delighted that yet again we have a rich and broad selection of papers to share."

Source: http://www.rarebirdalert.co.uk/v2/Content/WWT-pochard-decline.aspx?s_id=97764830

Promiscuous sandpiper flies 8,000 miles in search of mates

The male pectoral sandpiper (*Calidris melanotos*) has been known to fly 8,000 miles in search of a mate. Some males are more persistent than others. Researchers tracked one desperate small shorebird that logged more than 8,100 miles (13,045 kilometres) in two dozen different hook-up attempts over a frenetic four weeks.

"They're definitely trying hard to flirt and court," said biologist Bart Kempenaers of the Max Planck Institute for Ornithology in Germany . "They are not particularly successful most of them. Failed Don Juans mostly."

Sandpipers migrate from South America to

PICTORAL SANDPIPER.
Tringa maculata

breeding grounds in the Arctic tundra in the summer. The males tend to be sex crazy during this time because females are only fertile for a few weeks. They flit all over the place, trying hard to seal the deal with loud throaty hoots as many times as possible. The problem for them is that the females only mate once or twice a season. "Copulations are incredibly rare," Kempenaers said. "The males need to try and try and keep at it."

Source: http://www.rarebirdalert.co.uk/v2/ Content/Promiscuous-snadpiper-flies-8000-miles-in-search-of-mate.aspx? s_id=607364491

Asia's rarest seabird has been discovered breeding in the Korean Peninsula

A new breeding stronghold in the Korean Peninsula could help Asia's rarest seabird, the Chinese crested tern (*Thalasseus bernsteini*) pictured below in an image owned by Oregon Sate University, bounce back from near-extinction.

"Our hearts raced as we saw them," recalls Yunkyoung Lee and Se-Kyu Song. As part of a routine survey undertaken by the National Institute of Ecology of Korea, the team were on a rocky islet in the Yellow Sea, seven kilometres off the coast of southwestern South Korea. It was there that they spotted the species in the spring of 2016. "When we saw the distinctive headcrests we couldn't believe we were looking at two pairs of nesting Chinese Crested Tern", said Yunkyoung.

Only rediscovered 16 years ago on the east coast of mainland China, after its assumed extinction since 1937, only three breeding sites were known of this critically endangered tern, all on islands south of China. That is, until this

year: as well as a new site confirmed in the Taiwanese Strait, one chick has fledged from another, a South Korean colony all the way across the Yellow Sea.

"The return of Chinese Crested Tern as a breeding bird in the Yellow Sea is an extremely nice surprise," says Simba Chan, Senior Conservation Officer for Asia, BirdLife International. At other discovered breeding sites (Matsu Islands, Jiushan Islands, Wuzhishan Islands and the recently-confirmed Penghu Islands), Chinese crested tern are found breeding in mixed colonies of greater crested tern (*Thalasseus bergii*). Chan explains the importance of the new discovery amidst the gulls: "There are no breeding colonies of Greater Crested Tern north of the Yangtze estuary, so we previously thought that Chinese Crested Tern would not breed in the Yellow Sea region before their numbers reached a threshold large enough to form their own colony. The new site means the future of this species looks more promising now, as there are more colonies where it can nest."

Mike Crosby, Senior Conservation Officer, BirdLife International said, "Given this potentially larger range, we need to understand why the species is so rare – could it be caused by egg collecting and human disturbance at the nesting colonies?"

"We immediately requested the Ministry of Environment to restrict all civilian access (even researchers), and to secure the area until breeding success was confirmed," said Yunkyoung. "The Ministry also took action to help the birds."

Source: http://www.rarebirdalert.co.uk/v2/ Content/Asias-rarest-seabird-has-been-discovered-breeding-in-the-Korean-Peninsula.aspx?s_id=97764830

Royal tern's star turn

A first-winter royal tern (*Thalasseus maximus*) [pictured here by Nicholas Atamas, was discovered along the north-west coast of Guernsey in the Channel Islands. A record of this species anywhere in Europe in winter is of huge significance, and the usual questions regarding origin were quick to surface — African or American?

With one confirmed British record of American royal tern occurring at Kenfig Pool, Glamorgan, in late November 1979 (it had been ringed as a nestling in North Carolina the previous summer) and another ringed bird presumably having crossed the Atlantic to Mumbles, Gower, in December 1987, there appears to be a precedent – if only a small one - for transatlantic vagrancy in mid-winter.

SOURCE: http://www.rarebirdalert.co.uk/ v2/Content/WeeklyRoundup2017-06.aspx? s_id=975143597

African Penguins face perfect storm of climate change and over-fishing

Endangered penguins are foraging for food in the wrong places due to fishing and climate change, research led by the University of Exeter and the University of Cape Town has revealed.

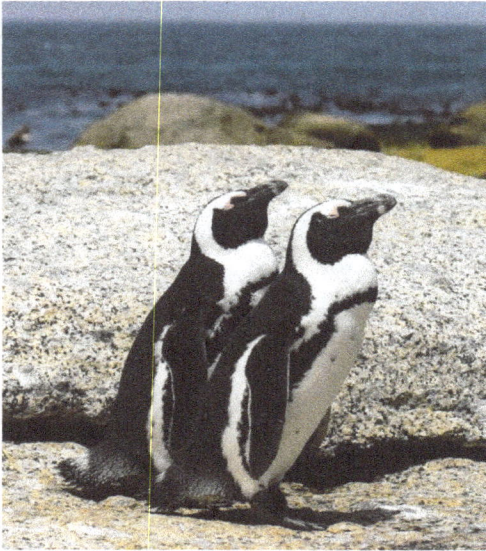

PIC: Charles Sharp

Juvenile African penguins search large areas of ocean for certain signs which usually mean there is plenty of prey, but rapid shifts caused by climate change and fishing mean these signs can now lead them to places where food is scarce - a so-called "ecological trap". The research reveals low survival rates among juvenile penguins, and models suggesting breeding numbers about 50% lower than if the birds were able to escape the trap.

"Environmental degradation can cause maladaptive habitat selection, meaning cues which used to work for a species now put them in danger," said first author Dr Richard Sherley, of the Environment and Sustainability Institute on the University of Exeter's Penryn Campus in Cornwall. "Juvenile African penguins look for areas of low sea temperatures and high chlorophyll-a, which indicates the presence of plankton and therefore the fish which feed on it. These were once reliable cues for prey-rich waters, but climate change and industrial fishing have

depleted forage fish stocks in this system."

Dr Sherley said overfishing off Namibia and the combined effect of commercial fishing and environmental changes off western South Africa had dramatically reduced populations of fish preyed on by penguins, and continued that small changes in the temperature and salinity of the waters in the area (known as the Benguela ecosystem) where fish such as sardines and anchovies used to aggregate had caused these species to move their distribution hundreds of kilometres to the east.

He added: "Climate change and fisheries are transforming the oceans, but we don't have a complete understanding of their impact. Our results support suspending fishing when prey biomass drops below certain levels, and suggest that mitigating marine ecological traps will require major conservation action." Dr Stephen Votier, also of the University of Exeter, said: "This ecological trap was only discovered when young penguins were tracked from multiple colonies. This highlights the power of studying animal movements, particularly for long-lived marine species like penguins. In fact, tracking is now a crucial tool in conservation biology."

Source: http://www.rarebirdalert.co.uk/v2/Content/Endangered-African-Penguins-face-perfect-storm-of-climate-change-and-over-fishing.aspx?s_id=97764830

British Barn Owls still struggling to adapt to modern life

The most recent nest site survey indicates that the factors that wiped out 70% of barn owls (*Tyto alba*) in the 20th Century still impact the species today.

By the mid-20th Century, changes to the dynamic of the human-bird relationship rapidly

sent British barn owl numbers into a nosedive. According to the Barn Owl Trust – a UK-based charity who works to protect the species – barn owls (as photographed here by Steve Garvie) in the country declined by as much as 70% between 1932 and 1985.

These declines are largely the result of improvements in the way farmers cultivate their land. The rise of the combine harvester, which is extremely efficient at harvesting grain, and the development of sealed grain silos, means grain is no longer stored in enclosures on farmland, meaning less food for rodents during the long, cold winters, and subsequently less prey for owls.

Today's farmers are also able to cultivate land that was previously beyond their tools' means, allowing them to plough right to the edges, resulting in the loss of the Barn Owl's favoured hunting habitat: rough grassland verges. Add to this, the 20th century has also brought with it other dangers that have taken their toll: road fatalities, potent rat poisons and the loss of nesting spots as traditional barns are pulled down and replaced with less inviting buildings.

It would seem that numbers of the species in the UK have stabilised since the mid-1990s. While numbers of the UK population of this species are not understood, the health of the country's barn owl population can be surveyed through the Barn Owl Trust's annual nest site survey. The number of nesting pairs in 2016 was down 6% on the all-year average, and the number of young in the nest was down 7%. While disappointing, the numbers are at least an improvement on 2013 and 2015, which were two extremely poor years where cold weather saw nesting occupancies down 70%

and 25% against the all-year average.

There is clearly a tendency towards low productivity, which the Barn Owl Trust attributes to a lack of available prey, and an overall low population density exacerbated by a lack of juvenile barn owls to replace the adults.

The report, which can be read at: http://www.barnowltrust.org.uk/barn-owl-facts/current-uk-barn-owl-population/ recommends several measures to help sustain Barn Owl numbers, including habitat improvement, the installation of low-flight prevention screens around trunk roads to avoid collisions, and replacing existing nest boxes to reduce chick mortality by ensuring the replacements are no less than 460mm deep.

Source: http://www.rarebirdalert.co.uk/v2/Content/Birdlife-British-Barn-Owls-still-struggling-to-adapt-to-modern-life.aspx?s_id=97764830

Farmers and conservationists securing the stone curlew's future

Farmers in the UK, together with the RSPB and Natural England, have helped secure the long-term recovery of the stone curlew (_Burhinus oedicnemus_). After a four-year EU LIFE+ project, this rare bird now has more safe nesting habitat away from crops, leading to hopes that their UK population will become sustainable within the next five years. Once widespread across farmland and heathland, stone curlew numbers crashed by 85% between the 1930s and 1980s due to habitat loss and changes in farming methods. Today, most of the breeding population is concentrated to small areas in the Brecks on the Norfolk/Suffolk border, and around

Salisbury Plain, Wiltshire.

To secure the future of these iconic birds, the RSPB and Natural England have been working with landowners and farmers to create more habitat suitable for nesting stone curlews. The £1.3 million 'Securing the Stone Curlew' project, funded by EU LIFE, began in 2012 with the aim of increasing the number of stone curlew nesting on safe ground to 300 pairs – equivalent to three quarters of the UK population. To do this, in the Brecks and in Wessex, south-west England, farmers have been creating 'fallow plots' on or near their fields which will lay undisturbed by machinery, allowing stone curlews to nest in peace. This also benefits rare plants, insects and other birds such as the threatened turtle dove which feed on the seeds from these plants.

Nearly 300 of these nest sites are now created by farmers each year, with support from stewardship schemes, and over 3,000 hectares of grassland habitat – the size of a small city – is now being restored to create the right conditions for stone curlews. With that, 144 more chicks fledged in 2015-16 compared to 2012-13.

Source: http://www.rarebirdalert.co.uk/v2/Content/RSPB-Farmers-and-conservationists-securing-the-Stone-Curlews-future.aspx?s_id=97764830

Slender-billed Curlew: have we been looking in the wrong places?

Scientists from the RSPB have carried out stable isotope analysis on feathers from juvenile slender-billed curlews (*Numenius tenuirostris*) to identify potential breeding areas, and found that these areas may be significantly further south than previously thought; previously the only known breeding area of this species was from the Omsk province, western Siberia. This has led some to hope that new searches could lead to its 'rediscovery'. The data from slender-billed curlew feathers from juvenile birds (donated from museums across the globe) and data from feather samples of other wader species from locations in Russia and Kazakhstan, collected by expeditions led by Geoff Hilton, now with WWT, was used to compare with large scale hydrogen isotope maps. A 'calibration equation' was produced which enabled the locations the juvenile

The **slender-billed curlew** (*Numenius tenuirostris*) is a bird in the wader family Scolopacidae. It breeds in marshes and peat bogs in the taiga of Siberia, and is migratory, formerly wintering in shallow freshwater habitats around the Mediterranean. This species has occurred as a vagrant in western Europe, the Canary Islands, the Azores, Oman, Canada and Japan. After a long period of steady decline, the slender-billed curlew is extremely rare, with only a minute and still declining population. This is thought to be fewer than 50 adult birds, with no more than two or three verified sightings in any year in the last five (as of 2007). As a result it is now listed as critically endangered. It is the first European bird species highly likely to become entirely extinct since the last great auk died in 1852.

Slender-billed Curlews grew their primary feathers to be identified. The results suggested that the breeding range may have been centred in the steppes of northern Kazakhstan and part of southern Russia, between 48°N and 56°N. This is considerably further south of the known breeding range and in a different habitat zone, being more in drier steppe than the forest steppe habitat in the Omsk area.

It seems possible that the records from the Omsk area may be atypical for the core range of the species and any future searches for the species may be better targeted at the Kazakh steppes. If it is extinct, the identification of the breeding sites may help understand the reasons for the species' decline and help learn wider conservation lessons.

Source: http://www.rarebirdalert.co.uk/v2/Content/Slender-billed-Curlew-have-we-been-looking-in-the-wrong-places.aspx?s_id=97764830

PIC: Andrew Shiva

they estimate a likely global population of 14 to 16 million. Before, population estimates only took into account breeding pairs, said Australian Antarctic Division seabird ecologist Louise Emmerson, who also explained, "Non-breeding birds are harder to count because they are out foraging at sea, rather than nesting in colonies on land. However, our study in East Antarctica has shown that non-breeding Adelie Penguins may be as, or more, abundant than the breeders. These birds are an important reservoir of future breeders and estimating their numbers ensures we better understand the entire population's foraging needs."

Source: http://www.rarebirdalert.co.uk/v2/Content/Adelie-Penguin-population-in-east-Antarctic-doubles.aspx?s_id=97764830

Footage of rare bluethroat filmed in Lincolnshire

A male bluethroat (*Luscinia svecica*) was spotted at Willow Tree Fen nature reserve in Bourne, Lincolnshire, and footage has emerged after a bird enthusiast filmed the bird during February.

According to the RSPB, between 85 and 600 of the birds pass through the UK during migration in the spring and the autumn, and the organisation says on its website: "Best looked for along the east coast in spring and autumn in scrub and grassy areas. Usually seen hopping along the ground or ducking into low cover."

Adelie Penguin population in east Antarctica double previous estimate

Almost six million Adelie penguins (*Pygoscelis adeliae*) are living in East Antarctica, more than double the number previously thought, scientists have announced, in findings that have implications for conservation.

Research by an Australian, French and Japanese team used aerial and ground surveys, tagging and resighting data and automated camera images over several breeding seasons, which allowed them to come up with the new figure.

They focused on a 5,000 kilometre (3,100 mile) stretch of coastline, estimating it was home to 5.9 million birds - some 3.6 million more than previously thought. On this basis,

The bluethroat is a small passerine bird that was formerly classed as a member of the thrush family Turdidae, but is now more generally considered to be an Old World flycatcher, Muscicapidae. It, and similar small European species, are often called chats.

It is a migratory insectivorous species breeding in wet birch wood or bushy swamp in Europe and Asia with a foothold in western Alaska. It nests in tussocks or low in dense bushes. It winters in north Africa and the Indian subcontinent.

Source and video: http:// www.lincolnshirelive.co.uk/footage-of-rare-bluethroat-bird-filmed-in-lincolnshire/story-30177070-detail/ story.html#DwQ4rVebvULrLB60.99

World's oldest wild bird just hatched another egg at 66

It was reported in February that the Laysan albatross (*Phoebastria immutabilis*) known as Wisdom, believed to own the title of the world's oldest known avian has hatched a new chick for the second straight year – no small feat for the approximately 66-year-old bird, as it takes approximately seven months to incubate an egg and raise a chick.

The US Fish and Wildlife Service (USFWS) revealed that Wisdom, who has been living at Hawaii's Midway Atoll National Wildlife Refuge and Battle of Midway National Memorial for more than six decades, had finally hatched an egg she that was first spotted incubating back in early December.

Wisdom, who according to NPR was first banded by biologist Chandler Robbins in 1956, has given birth to at least 30 to 35

chicks, said Bob Peyton, USFWS Service Project Leader for the Refuge and Memorial. When not incubating eggs, the agency said that she is typically extremely active, having flown an estimated three million miles over the course of her life. (Image courtesy Forest and Kim Starr).

Source: http://www.redorbit.com/news/ science/1113417462/worlds-oldest-wild-bird-just-hatched-another-egg-at-66/

For those of you not aware, as well as this column in *Animals & Men*, Corinna writes a daily Fortean bird blog which can be found as part of the CFZ Blog Network, but also as a stand alone site at:

http:// cfzwatcheroftheskies.blogspot.com/

Nederland

Noordzee

Wadden Zee

GRONINGEN

Delfzijl

Eems

Leeuwarden

FRIESLAND

Groningen

Smallingerland

Sneek

Assen

Heerenveen

DRENTHE

Den Helder

IJsselmeer

Emmen

NOORD HOLLAND

Hoorn

Meppel

Hoogeveen

Alkmaar

Purmerend

Zaanstad

Markermeer

Zwolle

FLEVO

Lelystad

LAND

OVERIJSSEL

Haarlem

AMSTERDAM

Almere

Almelo

Haarlemmermeer

Hengelo

Katwijk

Hilversum

Apeldoorn

Deventer

Enschede

Leiden

Amersfoort

Zutphen

Den Haag

Utrecht

GELDERLAND

Zoetermeer

UTRECHT

IJssel

Westland

ZUID

Lek

Ede

Arnhem

Delft

Rotterdam

Nederrijn

Doetinchem

Hellevoetsluis

HOLLAND

Merwede

Waal

Rijn

Haringvliet

Dordrecht

Oss

Nijmegen

Grevelingen

's-Hertogenbosch

Zierikzee

Maas

DUITSLAND

Oosterschelde

NOORD BRABANT

Roosendaal

Breda

Tilburg

Helmond

Middelburg

ZEELAND

Vlissingen

Bergen op zoom

Eindhoven

Wester schelde

Venlo

Terneuzen

LIMBURG

Antwerpen

Roermond

Schelde

Maas

BELGIE

Geleen

Maastricht

Heerlen

Provinciale hoofdsteden in rood

A wikipedia Map

BLACK DOGS IN THE NETHERLANDS

Writing about tradition is a wonderful thing. When I started investigation literature in search of Dutch Black Dog traditions I never would have thought there is so much on this subject in my far from mysterious country.

My attention has always been on the English and American tradition, and never on my own.

That changed when I came in possession of a book of **Marten Douwes Teenstra**, *Nederlandse Volksverhalen*, published in **1843**. Teenstra had one objective: to battle superstition in his home province of Groningen. In order to bring this wayward province into the 19th century, Teenstra believed people should get educated. Superstition was not helping that process. With that in mind Teenstra described extensively everything people believed in, providing us with a wealth of data on the shadow world, including black dog lore.

Groningen is one of our most northern provinces, flat as most of our country, windswept, with a long coastline and endless vistas of pastures and sky, where even now it is imaginable that strange creatures roam the lonely landscape.

Familiar as I am with the Black Shuck lore of the UK, I was surprised to find black dogs in Groningen.

They didn't belong there. Even more surprised I was when my research ended up with black dogs all over our country, and also in Germany. Here's the language barrier at work, for as far as I know from my own books none of the writers on the Black Shuck jumped the North Sea.

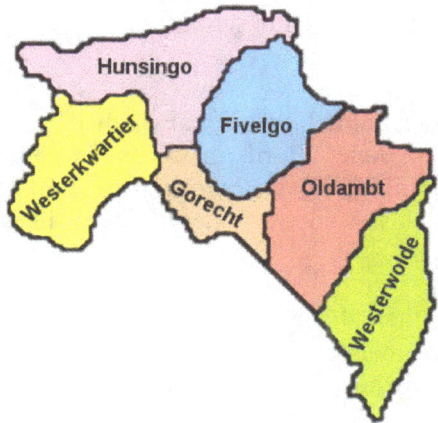

LOES MODDERMANN

55

So: Meet the **Börries.**

Also known as **'Stommelstaart'.**
'Plaagbeest' or **'Helhond'.**

The physical description of the Börries doesn't deviate much from the well known one. The beast is said to be shaggy, big, with extremely large, red fiery eyes, 'a big as dishes'. And like the English Shuck it appears in lonely places, makes (in most cases) no sound, is seen without a head, grows from small to big or changes into another animal, and sometimes drags chains. It appears and disappears at will, leaving the witness in turmoil.

Sightings of the Börries are linked to the devil, and also often to pending disaster or death.

A difference between the Black Shuck and the Börries is, maybe, that the Börries is often described as a 'poedel', a poodle.

I don't know if this is really a distinction, since Janet Bord mentions in *Alien Animals* (p79) that the coat of a Black Shuck is sometimes described as 'like a sheep'.

Another thing is the tail. The name 'Stommelstaart' refers to no tail at all. Meanwhile Teenstra and others point out that the Börries has a rough tail standing stiffly from the body.

Several witnesses also mention the ambling gait of the dog: stiff, quite unlike a normal dog.

On the illustration one of the many renderings of the Beast of Gevaudan, which has some similarity with a curly 'poodle' Börries, except for the tail...

Often the Börries is seen near water, ditches, or on river dykes, sometimes on a certain stretch that is regarded as haunted.

Like his Angelsaxian brother he is also connected to crossroads. I know nothing about possible Leylines in Holland, as far as I know there has never been research done on the subject here. Maybe they do exist, maybe not.

An other place where he is often seen is on *'wierden'* (or *'woerden'*), an old name for former gravesides and a word that is found in a variety of Dutch place names. It is not surprising that also cemeteries and execution spots were haunted by the Börries.

The stories as they are taped in the 20th century are understandably scetchy, almost always hear-say from long ago. When Teenstra wrote down his investigation in the mid-19th century, these stories were very much part of daily life. He is thorough: naming many places in Groningen where the Börries makes a well defined appearance.

So is the Börries in *Oostwold*, in Westerkwartier seen with a foal without a head and sometimes a calf with a chain around its neck. But near _Tolbert_ and _Marum_, in the same region, it is seen in the company of a flying pig. _Marum_ - mare is an old word for water.

On September 24 1731, 22 young man, the youngest 15, were convicted, killed and burned in _Faan_ (Fanum) for sodomy, or rather homosexuality, and six more tortured and locked up. Since then this region was plagued by some rough Black Dogs with fiery eyes, pulling an iron cart (wherewith the convicted men were driven to their death?) which made a terrible squeeking noise.

The wild hunt isn't unknown in these parts either. Teenstra mentions that he is seen with his hounds in _Zevenhuizen,_ also in Westerkwartier.

He mentions several cases of the Börries in the guise of a 'waterdog' , strolling along the dykes, and even on the ice in winter. Between midnight and 4 a.m it's foolish to be outside, for these are the hours the Börries is most active.

In several places in Hunsingo people see the Börries walking certain paths, sometimes following people but when they stand it stands too.

In _Warffum,_ Hunsingo, there are several helhounds walking around a convent, pulling fiery chains, and in _Rottum,_ also in Hunsingo there are 4 or 6 hounds pulling a fiery cart, back and forth a lane of the former cloister of Rottum.

I found one story about the 'Widde Wiend', (below) a Black Dog in the shape of a Greyhound who grows and 'is obviously the devil' . He specifically haunts het 'Hooge Land', a region NW of the city of Groningen. Looks innocent until you don't follow him any longer. (*K. Ter Laan, Groninger Overleveringen* I)

Some of these stories are difficult to interprete. One wonders how many totally natural dogs in mysterious places, at night, were labeled Devildogs.

Some stories seem to confirm that. The paranormal abilities of dogs, and their premonitions of death were perhaps interpreted als 'devilish' .Dogs were no pets, they were guard dogs or they roamed the country side. Rabies was a very real danger. In places where the fear for the Börries was rife, every black dog could be one. That fear is also confirmed in the many other black dog stories I found in other parts of the country.

Teenstra mentions Black Dog traditions in **Friesland,** the Province west of Groningen. He doesn't elaborate, he only talks about Groningen. Friesland has her own culture and her own language that's related to English.

On internet I found lots of Frisian stories of Black Dogs, that are not called Börries here, but mostly 'Spookhonden', Spukdogs.

The following stories are almost all from the Verhalenbank of the *Meertensinstituut*, Amsterdam. This institute was founded to preserve Dutch language and traditions in books and collections, paper and digital. I don't think there has ever been a scholarly investigation in Black Dog lore. They are hidden in books and tapes and written reports, most of them from at least half a century ago. Most stories are not dated, the only date is when they were recorded, which can be long after they actually went around in a village or region. I picked a few from the files of every province, but there is much more to find.

The stories of Black Dogs in Friesland are not essentially different from the Groninger Börries. The apparitions are often connected to the devil and to a bad omen. Several stories mention a Black Dog looking someone in the eyes, conveying the notion of death. Soon thereafter this person, or someone of his houshold gets sick and dies.

Like a black dog with glowing eyes in *Tjietjerksteradeel* who follows a woman home, and then disappears.

Or in *Zwagerbosch* around 1910 where some people saw a black dog going through a wall.

In *Drachten* (1954) father and son were on their way to a sick grandfather when they met a black dog in the field. That has to be the devil, they thought. Then they met Kornelis Wijtses, professional devil bannisher, who banned the devil. The dog was not seen again.

There's no mention in this story about a relationship between seeing the dog and the sickness and probably dying of the grandfather.

Another province that's rich in Black Shuck stories is **Zeeland,** in the southwest of the country, a province that is mostly water and islands, and has had a tough time fighting the floods in the past, before powerful dykes protected the land. Many are the legends about drowned villages where the bells are still heard ringing under the sea.

In Zeeland the Black Dog is called **'Ossaert'.**

There's one interesting and also funny story from around the 18th century. Zeeland became Protestant after the Reformation (it still is), but the new brand of religion couldn't suppress Catholicism completely.

In Zeeuws Vlaanderen, the southernmost part of the province, adjacent to Belgium, there was the small town of *IJzendijke,* where a fanatical vicar wanted the monks of a convent in IJzendijke gone. They went to the town of *Sas van Gent,* but not willingly.

Five years later the vicar was smitten with a deadly disease, of which he died and was placed on a bier.

Then an Ossaert arrived chasing away all the people, making it very clear that the vicar had led a less than exemplary life. Nobody could, or dared, to chase the beast away, but all agreed it was the devil himself.

They sent for another vicar from a neigboring village to get rid of the Ossaert, but the Bible didn't make any difference. The dog spit fire and stayed where he was.

There was only one way out of this deadlock: ask the monks from *Sas van Gent* to help., the ones that were driven out of their belongings by the dead vicar. They came ...and they brought a consacrated hostie.

That was too much for the Ossaert, and he jumped through the window in fire and flames. The Monks got their possessions back in IJzendijke.

This story, still part of Zeeland's folklore, shows how much the Black Dog in some places is mingled with religion and with the devil.

Overijssel is another province with a Black Dog tradition.

Many stories were told about the spukdog, also named **'Kardoes'.**

He even got his own statue in the village of *De Lutte,* made by sculptor Pieter de Monchy, after the Egyptian deity Anubis, the protector of graves and souls.

A fragment from an article in de **Telegraaf of 15-05, 1993,** tells what people believed. Translated:

DE HELHOND VAN
DE LUTTE

"In the vilage of De Lutte near Enschede stands a brazen statue of a large disastrous dog. It's the image of the ghostdog who used to make the region unsafe in the past.

The people of De Lutte imagined that the dog was one of the barking animals that accompanied Wodan on his wild hunt. The story went that when the helhound barked somewhere there would soon be a death. Apart from in the vicinity of farms the dog could also to be found on cemeteries.

The story attached to this helhound is one of several where being in contact with a ghostdog is connected with death or illness. In other words: where the dog appears, there is not much people can do to escape fate.

The province of **Gelderland**, a large region with big and smaller rivers, small hills, a big sandy area (*Veluwe)* and lots of wood, has a long tradition of folk tales, Black Dogs among them.

Here's one from the village of *Gendt*, at the river Waal, near the German border.

A man returns home from a wedding in Germany, walking alone. He had too much to drink. A black dog joins him and walks alongside, till he is home. The animal doesn't leave, so he tells it that if it is the devil, it has no power over him. The dog begins to grow. The man makes the sign of the cross and the dog leaves.

In *Lopik* a man is buried. A black dog walks behind the funeral procession. When everyone is gone the dog lies himself on the grave and stays there. The dog was not the man's, no one had seen it before. The dog keeps lying there till it dies.

Sounds like a real dog to me!

In *Hoog Soeren* there was the so called *'Juffersboom'* , a large and thick hollow tree where a white lady was spinning and black dogs were seen around the tree. People were afraid to go near that tree.

Once a man went to Apeldoorn, the nearest city, where he visited the pub and bragged about the white lady negatively. Walking back in utter darkness (even today it's very dark and lonely there) he came upon the Jufferboom. There he saw a black dog with fiery eyes disappear into the woods. Then an invisible thing attacked him, slapping him around the ears and knocking him down. There was nothing. The man tried to defend himself but the attacker was invisible. In the end he came home, very much in shock.

A man in *Liemers (*1970) walks on a path along a churchyard. A black dog suddenly appears through the hedge around the cemetery. The man is not afraid and strokes the animal, who suddenly starts to grow until it is higher than the man. Then it disappears.

All in all the belief in witches in Gelderland was stronger, I think, than the belief in the Black Dog. There are multiple stories about black cats and shapeshifting witches.

I searched in every province and it turns out that the Black Dog is universal, although some provinces have more on them than others, which is probably due to the landscape and the early expension of cities. A few provinces, **Drente and Utrecht** have almost no material, at least that I can find. That isn't necessarily true. Often such stories are very local and never turn up in accessible collections. In a Drentse newspaper there was mention in 1949 of a black dog in *Ees* who was often seen. One day he panicked the horses of a young farmer's cart. The man was thrown and died. After that the black dog was not seen again.

There was a belief in **Zuidholland** that a ghost dog was the spirit of a dead person, so you had to take care not to hurt him. Other stories convey just fear of the devil, and like a black cat a black dog was a 'certain' creature from hell. Poor animals!

In *Alphen aan de Rijn* , Zuidholland, a night worker found a black dog on his way which he couldn't pass. He started praying and the dog disappeared.

In *Zoetermeer* a signalman found a big black dog, as big as a calf quite nearby while doing his job at 9 p.m.
The beast crawled under a fence and disappeared in the polder. His daughter saw the huge animal too.

The next day the signalman inspected the fence where the dog went through, but there was no way he could have done that, because there was not enough space for even a small dog.

In a paper from 1897 I found a case about a guard soldier standing outside in *Woudrichem* when he saw a black dog .

For target practice he tried to shoot it but it jumped 3 m in the air and disappered. This black dog was seen more often in *Woudrichem*, and called a werewolf.

In *Uitdam* in **Noordholland** (1899) there was a farmer in a flat boat going under a bridge when he saw a black dog with fiery eyes jumping. He missed the boat and called out: "That's your luck!"

Those black dogs were thought to be the ghosts of criminals buried under the scaffold in that village.

As for the province of **Noord-Brabant**, there's a lot. Some stories mix werewolves with the Black Dog.

Like the story of a man in *Haaren* who changed into a big black dog at night who jumped on your shoulders.

In *Middelbeers* the devil changed himself into a big black dog . People didn't dare to go to church because that dog was on the steps of the church and stayed there, showing his teeth. Then the pastor, a courageous man who saw his church become empty, hit the dog on his behind, telling him to get lost in the name of the Lord. The dog spoke: " All right my wife and children, I'm off" , and ran away howling and yelping. He never came back.

It was clear this was a man who was a notable, but recently deceased blasphemer who took the appearance of a black dog to see his wife and children going to church. Blaspheming attracts the devil, a worthy lesson for churchgoing people in those days.

Someone tells the story that his grandfather once came home and found a big black dog in his barn. He chased him out, but shortly after all his 12 horses got ill and died. The vet didn't find a cause, and the horses looked normal.

A bit more 'classical' Black Dog story is the following from 1889, *Oerle* near *Veldhoven* in Noord Brabant.

On a winterevening a farmer from *Oerle* went along a footpath that was divided from a broad sandroad by a hedge.

Suddenly he saw on the other side of the hedge a big black dog that walked along with him. When the man halted, the dog halted too. The farmer knew this place was said to be haunted and he understood what the dog really was. But not being afraid he broke of thick branch to defend himself.

The dog in the meantime had grown considerably and was now a big as a foal. They never met: before the farmer came to the place where the footpath became a bigger road, the ghostdog retreated and disappeared.

It seems this was a dog who was bound to a certain stretch of the road and couldn't go further.

A story from Brabant, 1968, place unknown, where the werewolf and the Black Shuck are one and the same.
A man tells about his father who met a werewolf who came out of a ditch in the shape of a black dog, and went into another ditch, passing the father. Such things happened to people who cursed a lot.

The father was such a person.

The last one from Noord Brabant, *Eersel,* during the Belgian Revolution (1830), when Dutch troups camped at the border with Belgium. There was a captain who mistreated his men and controlled them harshly every night.

He was very much hated by all the soldiers. He then disappeared mysteriously. But after that a big Black Dog came every night with a clanking chain to check on all the posts, and nobody dared scare him away. Some tried to shoot him but the dog seemed invulnarable for bullets. When the soldiers left the border, the dog also disappeared.

Limburg, the most southern province of the Netherlands, bordering Belgium and

Germany, has classical Black Shuck stories:

In the region of *Venray* there's a black dog with fiery eyes and a big heavy fiery chain around its neck. It wasn't agressive, just lying around 10 p.m. near a friar convent, with the chains over the road. People coming and going could easily step over the chains. In Venray the dog was called the **friar-dog**.

In *Neerpelt*, just over the Belgian border, a black dog with glowing eyes was seen very often. He was bound with a chain, on a narrow strach of the road. This dog was the devil. When the road was modernised, the dog disappeared.

A very peculiar story from *Meerssen*, involving a werewolf is the following:

A young girl is staying with her aunt. In the night the window of her bedroom is opened from the outside.

A black dog appears outside, behaving crazy, sommersaulting and dancing, standing on its head, and when disappearing it sticks out its tongue and bows mockingly to the girl and her aunt.

The father of the girl's fiancé is said to be a werewolf. It must have been him.

In *Swalmen* (1962) there was a house where once lived a butcher with a bad character. When he died a black dog was seen around the house, and a fiery handprint was on the window, which couldn't be removed.

In *Stamproy* A Black Dog was seen multiple times on the cemetery. When people came near it grew bigger and bigger, also his eyes, and then it disappeared.

This one from 1925 is special, fitting into the very religious, Catholic tradition of this province. In *Gulpen* there lived a very bad man. Every time he went home a black dog went with him. That dog grew with the amount of his misdeeds. One day it changed into a big black horse, chasing the man, who became ill. The next day a Black Dog sneeked in and went under his bed. It couldn't be removed or chased away.

Every night it tortured the man who couldn't defend himself against the devil.
The priest was called and he advised the man to say confession. He did, but the dog stayed.

Only the ninth time he said confession the dog came from under the bed, yelping, and jumping through the wall, leaving the stench of sulpher behind. That hole in the wall could not be repaired, until a stone was cut and blessed and inscribed with a cross.

The man got better. If he mended his ways the story doesn't tell...

Conclusion: Black Dog stories are varied in this country, and in many cases they were not supernatural at all.

But an otherworldly interpretation says something about how deeply ingrained the belief in the supernatural was in rural regions until fairly recently. We no longer see the devil in a black dog or cat. The question is, did the Börries and his family really disappear, or are they just in hiding?

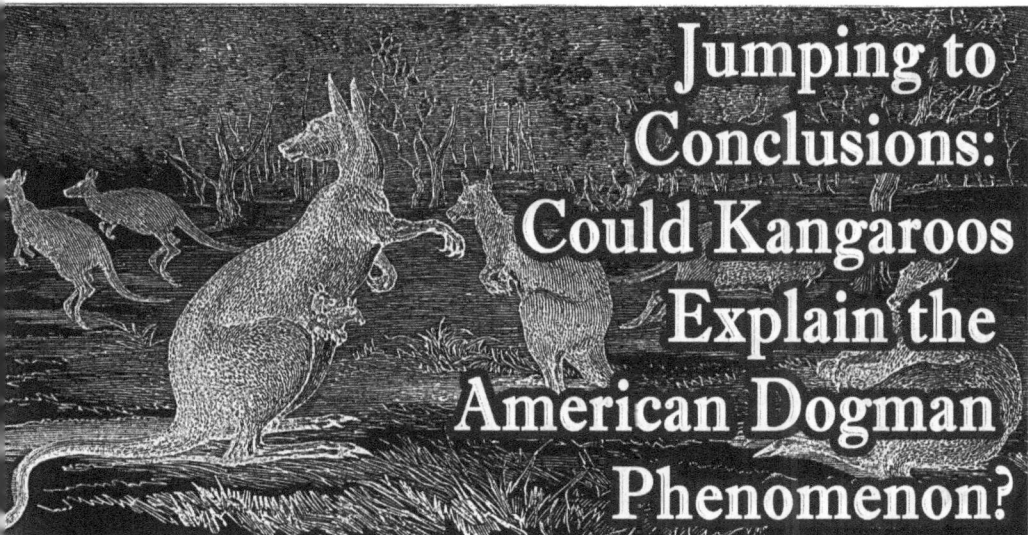

Jumping to Conclusions: Could Kangaroos Explain the American Dogman Phenomenon?

Since the mid 1990's, the idea that bipedal wolves or dogs roam the American wilderness has exploded. Attention to the phenomenon began with the sightings of a bipedal canid creature that was encountered by numerous people along a farm-road near Elkhorn, Wisconsin. The creature, dubbed the Beast of Bray Road (after the road it was seen along), became a media sensation and propelled then small town journalist Linda Godfrey into near stardom in the cryptozoological field. Since then, the Dogman has been reported in nearly every state in the country. The amazing amount of attention that the creature aspired also drew many new researchers to the field and lead to the creation of several nationwide groups, such as the North American Dogman Project lead by Ohio based cryptozoologist Joedy Cook.

The Dogman is a fickle thing. Nearly every sighting is of a physical creature – literally a bipedal canine – but, as of right now, there are no known evolutionary parameters or niches that would be filled with the existence of such a creature. That leads some researchers, such as myself, to look for other viable options, especially in the known animal kingdom. This method of investigation took me to a conclusion that might seem a little extreme at first, but I feel is a viable possibility for some sightings of the captivating mystery of the Dogman.

Those Hoppy Things from Down Under

The kangaroo is among the most beloved exotic animals in America. The kangaroo can be seen in everything from ad campaigns to candy to stuffed toys, and has been for many years. Sightings of out-of-place kangaroos are nothing new as well. Phantom kangaroos have been seen in America since at least the 1970's, during which a wave of sightings exploded in the Midwest. The best example of a sighting during the aforementioned wave occurred on

COLIN SCHNEIDER

October 18th, 1974. Two Chicago police officers, Leonard Ciagi and Michael Byrne, were patrolling the Northwestern side of the city when they received a report that a kangaroo had been spotted. When they went to investigate, they found and followed a trail which lead them to the kangaroo. The officers found the kangaroo at the end of a darkened alley. The animal was described as about 5-feet-tall, hunched over, and was visually agitated. When the officers tried to detain the marsupial – by handcuffing the animal – it reacted by growling and screaming. Ciagi was kicked in the leg by the thrashing animal. They then decided to back off until backup came. When help did arrive, the kangaroo made its escape, bounding over a fence and away from the onlookers.

The Ciagi and Byrne sighting isn't unique; during the 1970's kangaroo wave; growling and aggressive kangaroos were fairly commonplace among the sightings. This isn't too dissimilar to how kangaroos act in the wild. The common red kangaroo – the largest species of kangaroo in the world – is known to be fairly aggressive when approached by humans and has been known to attack passersby. The growling or grunting sound is also common with any species of kangaroo – the screaming, though, is not. Many assumed that the patrolmen's sighting was simply that of an escaped kangaroo, yet there is no evidence that there was any kangaroo that had escaped within 50 miles.

What about the Dogman?
Common descriptions of the Dogman do resemble that of an out-of-place kangaroo. Dogman witnesses often report seeing a bipedal animal with a long tail (which sometimes drags on the ground), pointed, upright ears, and a long, drawn-out face. Many witnesses also describe arms that reach out with clawed hands that hang from the arms and the creature standing in a hunched over position. All of these descriptions fit the kangaroo quite well – although they also fit the theoretical physiology of an upright canine. Let us then examine a case that could support the idea.

According to Linda Godfrey's encyclopedic work *Real Wolfmen: True Encounters in Modern America*, in the mid 2000's a young man (who wished to remain anonymous) was driving through Siloam Springs, Arkansas. As he was about the cross a bypass over some train tracks, he claimed to see a large, bipedal creature leap over the rails, onto the bypass. The creature was described as having a head similar to a jackal's or a greyhound's and had its forearms "'curled up' next to its body…" The creature crossed the bypass in several bounding leaps and vanished into the darkness.

The witness, who is Native American, felt that what he saw was a shapeshifting creature. The description, though, fits quite well with a kangaroo. Unfortunately, since it is unknown exactly when the encounter took place, the escaped kangaroo theory is mute. I couldn't find any evidence of an escaped kangaroo, or even a pet kangaroo in the area around that time, either.

How Plausible is this?

The idea that a species of kangaroo is roaming the wilderness of America is even more ludicrous than a bipedal canine. Modern kangaroos evolved to specifically fit the biological niches that were found in prehistoric Australia. That being said, there are non-marsupials that have evolved to have hopping as their primary locomotion through convergent evolution; the kangaroo rat is an excellent example. Yet, there is no known large mammal that has developed the kangaroo's predilection for hopping. What this boils down to is what the chances are that a large unknown hopping mammal is leaping across the country versus the idea of a bipedal canine lurking in the shadows of the country. Neither idea is especially convincing either way, yet there are hundreds of witnesses that have seen both; many in the same areas as the others.

What about zoo or pet escapes? The idea that cryptids could be misidentifications of known animals that have escaped from private collections or local zoos is one that has been flaunted around by skeptics for years, yet an actual escape is rare. It certainly does happen, but there is nowhere near as many escapes as there are sightings of unknown animals. The possibility is certainly something that should be looked at but will likely turn up nothing.

As unlikely as a bipedal dog is, it certainly is still the best biological explanation we have as of now. As I have shown, there are definitely cases where the Dogman that was supposedly encountered could plausibly be an escaped kangaroo – and the possibility should be investigated for every possible Dogman encounter.

Bibliography

- Coleman, Loren. *Mysterious America.* Winchester, MA: Faber & Faber Inc., 1983. Print.
- Clutton-Brock, Juliet, ed. *Mammals.* London: DK London, 2002. Print. Smithsonian Handbooks.
- Frances, Peter, and Angeles Gavira Guerrero, eds. Animal Life. New York: DK, 2008. Print.
- Godfrey, Linda S. *American Monsters.* New York: Jeremy P. Tarcher/Penguin, 2014. Print.
- Godfrey, Linda S. *Real Wolfmen: True Encounters in Modern America.* New York: Jeremy P. Tarcher/Penguin, 2012. Print.
- Godfrey, Linda S. *The Michigan Dogman: Werewolves and Other Unknown Canines across the U.S.A.* Eau Claire, WI: Unexplained Research Publishing Company, 2010. Print.
- Houck, Tyler. *Cryptid U.S.: Tales of Bigfoot, Lake Monsters, and More across America.* N.p.: CreateSpace Independent Publishing Platform, 2015. Print.

EDITOR'S NOTE: Whilst looking for illustrations for Colin's article I came across this short comic strip by Scott Kroll, which is reproduced with his permission: http://kingcrowcomics.blogspot.co.uk/

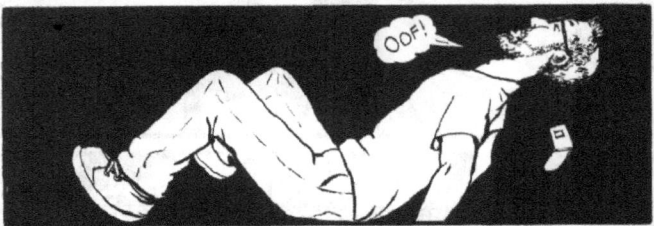

STILL ON THE TRAIL OF THE TASMANIAN WOLF

January of 2017 saw my third trip to Tasmania in search of the iconic flesh-eating marsupial known as the thylacine or Tasmanian wolf. I slept for most of the thirty-one-hour journey. Luckily, like Ralph Wiggum, sleep is where I'm a Viking.

I was once again teaming up with Mike Williams from the CFZ's Australian office. We found that having fewer people on the expeditions was easier in terms of organization.

We also made less noise in the bush. Again, some people have requested anonymity and we must respect that.

Once more Toyota kindly assisted us by providing a car and fuel card that gave us unlimited petrol. This took a major financial burden away from us and allowed us to move right across the island.

On the first day Mike had arranged an

RICHARD FREEMAN

interview with a journalist in Launceston. We visited the offices of *The Advocate* and Mike spoke with the journalist involved. I have an inherent mistrust of media types who have misrepresented me, and my colleagues, in the past. Therefore I sat the interview out. Mike hoped that a newspaper story would turn up new witnesses and give us leads.

Later, we met up with Grenville Batty. Grenville was the nephew of Wilf Batty, the last man to shoot a wild thylacine on 6th of May 1930 at Mawbanna, a hamlet in the north west of Tasmania. He once believed that his family still had the gun that did the awful deed but apparently, he has since found out that the gun he has is not of an old enough make to have been one in the same as his uncle's.

Grenville is an amiable chap in his early 70s. He has a keen interest in the Tasmanian wolf and thinks the animal may still exist. I had met and talked with him on a previous trip. Grenville had told Mike of a farm where possible thylacine activity had taken place. In the 1990s the place was a sheep farm. The owner had some predator that was taking out sheep on a regular basis. The kills were quite distinct from the dog attacks. At the same time the distinctive thylacine call, a high pitched triple yap, was heard. One of the workers reported seeing a thylacine from a car in broad daylight. The farmer switched from rearing sheep to rearing cattle and the kills stopped. The calls however were still heard by the farmer and his workers.

Then just four years ago, the farmer himself had his own sighting. As with his employee the animal was seen from a car at the side of the road. It was lying in vegetation and at his approach stood up and walked calmly away. He got to within five meters of the thylacine.

We had permission to camp on the farmland so we drove up and set up some baited trail cams. We made camp then went for a night drive. As with previous trips we utilized a crash cam, a windscreen mounted miniature camera that is constantly recording. We saw Bennett's wallabies, red bellied pademelons and a short beaked echidna. By the time we returned to camp it had started to rain heavily.

The following day we met up with Grenville again. He told us that whilst trapping possums in 1963 in the area, he had accidentally trapped a huge tiger quoll or spotted quoll. These are flesh eating marsupials related to the Tasmanian wolf. They have dog like faces and bodies not unlike a cat but somewhat more robust. They are a fawn colour with distinctive cream coloured spots. The tiger quoll is generally the size of a large domestic cat at around seven lb. The creature Grenville had caught was much larger, comparable in size to an Australian cattle dog a breed that weighs 33 to 49 lbs, similar in size to a border collie. Grenville indicated the height of the animal with his hands at around 18 inches.

This is not the first account of a giant-sized tiger quoll we have come across. On my first trip to Tasmania we interviewed a man who ran a market garden. We were mainly talking about a thylacine sighting he had in the 1990s but he also told us that back in the 1960s he had shot a huge quoll. The animal had been killing his chickens. He said it was the size of a cattle dog with a thick bull neck. He raised his hand to indicate the length of the creature when hung up. It was around five feet long. We heard of another of similar size seen the Cradle Mountain area.

Grenville wanted to give the animal to a local zoo but was too scared to tackle the angry beast. He was only fifteen at the time. He told the farmer who's land it was on and intimated that he should give it to the zoo. In an act of

wanton cruelty, the farmer threw the quoll into a barn with four dogs that ripped it to shreds. Apparently, the dogs were also badly injured by the giant quoll's teeth and claws.

It is possible that these outsized quolls represent a new species or, more likely. Freakishly large individuals of the tiger quoll. The Queensland tiger is a cat-like marsupial the size of a puma reported from the tropical north of Australia. Some cryptozoologists have postulated that it could be a surviving form of *Thylacoleo carnifex,* a long extinct flesh eating marsupial distantly related to the wombat. Perhaps some huge form of quoll is a more likely candidate is such a creature exists.

Grenville had also mention that he had a 'shack' in the area that we could use. Despite the fact that he told us that it had no electricity we jumped at the chance. Mike had been

sleeping in the car and I in a tiny, one man tent. I loathe camping despite having to do it at length on almost every expedition I've been on. I seldom get a good night's sleep under canvas, finding it cramped, cold and uncomfortable.

Grenville took us to the aforementioned shack. Mike and I were expecting some malodorous shanty one step from a garden shed. In fact, the 'shack' was a four-bedroomed farmhouse with a kitchen, living room, shower and toilets. The only reason that it had no electricity is that Grenville had not yet turned it on. One flick and we had light and heating. We used the farm as a base whilst we stayed in the north.
After settling in we did a night drive seeing wallabies, possums and Tasmanian devils.

The one minor down side of our Toyota sponsorship was that the fuel car could only be used with one specific company. All the garages in towns nearby had changed since our last trip and now the nearest garage of the company in question was many miles away in Wynyard. We brought some large fuel containers to stock up and save wasting time on too many long runs for petrol.

In a small museum in Wynyard we came across a pamphlet 'The Tasmanian Tiger Trail' by Colin Berry. It contained accounts of a number of sightings, some of which I had never heard. The publication had once come with a DVD that apparently contained an interview with the late Wilf Batty. The museum didn't have a copy but Mike wanted to try and track it down.

During a night drive, we took a wrong turn on the overgrown, labyrinthine tracks in the wilderness. We spent the better part of an hour lost and had to stop the car several times to pull logs, fallen trees and branches out of our way. We finally found the right path and made our way back.

Next day we returned to Wynyard for a meeting with a councillor who was, many years ago, involved in the publication of the thylacine pamphlet and the accompanying DVD. He could not find the disc in his archives and was sure that it was actually just an audio rather than film.

We met up with a man who had seen a thylacine on the farmland we had camped on. In 1990 he had been driving along a wooded road on the property with two passengers in his car. All of them saw a Tasmanian wolf emerge from the vegetation at the side of the road. It seemed like it had been resting there and got up when it heard the car approach. The animal looked at them before walking off into the forest. All of them had a clear and good view of the animal in broad daylight.

Mike's interview had begun to attract callers with their own stories. One man, who claimed to be an experienced hunter, said he saw a thylacine in Queensland on mainland Australia many years ago. He said that he and some friends saw it emerge from some bushes. His description of the animal however, lacking stripes and having a moth-eaten look, was that of a dingo with mange not a Tasmanian wolf.

We collected our camera traps. On the way to get one of them I heard a high-pitched yip. I froze in my tracks straining to hear. Nothing came. I move on again and once more the yip sounded. It was a single yip, not the triple yip associated with the thylacine but it still intrigued me. As I moved once more the yip was heard again, then I realized it was my boots squeaking! We checked the cameras. They showed Tasmanian devils, wallabies and quolls.

We drove down to the tiny village of Corinna. The place is named after the Tasmanian Aboriginal name for the thylacine. The drive was a long one through very wild territory on poor roads. The village itself is tiny consisting of a small pub come restaurant and a handful of houses. It is beside the Pieman River and was a former mining colony. Unfortunately, the ferry across the river was broken and we could go no further.

We moved on the Derwent Bridge and Mike had another contact, this one more promising than the last. The sighting had occurred when the witness was a boy in 1951. It was in the Central Highlands and at five in the morning when the witness was travelling with his father, a farmer. They saw a pair of eyes on the dark road ahead and thought it was a calf that had wandered out onto the road. As they drew alongside the creature the boy looked at it from a distance of only one meter from the car's window. It was a dog-like animal with striped hind-quarters and a long, stiff looking tail.

The boy was later interviewed by Dr Eric Guiler, a zoologist from the University of Tasmania. Guiler dedicated much of his life to researching the Tasmania wolf. Guiler apparently believed that the boy and his father had indeed seen a thylacine.

Later we moved down to Derwent Bridge on the edge of the Franklin Gordon National Park.

A man called Roy told us of a man he knew that had an odd encounter in Wuthering Heights, a coastal plain near the Frankland River in north west Tasmania. Some years ago, the witness, a logger had been in camp with several friends. They heard a yip-yip-yip vocalization and a crashing in the undergrowth. Suddenly a wallaby exploded from the undergrowth and ran towards the men. The

animal seemed exhausted and was panting. It actually hid behind the informant's legs. It was obviously very scared but the men never saw what was chasing it. He thought that it was being hunted by a thylacine.

More night drives revealed lots of wallabies but no devils or quolls. It seemed that these predators were lacking from this particular area.

We met up with Col Bailey, who I had been introduced to on the last trip. Col is unquestionably the greatest living thylacine hunter and as well as stalking the beast through the wilderness (an endeavour that rewarded him with a sighting in the 1990s) he also interviewed old bushmen, loggers, prospectors and hunters back in the 50s and 60s. These

people had first-hand experience of thylacines in the Tasmanian wilderness during their official period of existence. They are now all dead and without Col efforts their knowledge and stories would be lost to the ages.

Col told us about his expedition into the far south west of Tasmania. The South West Conservation area is an uninhabited section of the island. A wilderness of windswept button grass it is seldom trodden by man and has only one rough track leading into it from an arm of the Macquarie Harbour, a large, shallow, natural inlet. The track peters out after a short while.

Col originally planned to go to the area by boat but the swell was too great in the Southern Ocean. He ended up chartering a helicopter to drop him on the west coast of the area, an endeavour that cost him $5000 and today would be far, far more costly. Col spent a week at the ends of the earth trekking up and down the coast and venturing inland. He found possible tracks on a remote beach. He almost ran out of water and found that the creek near his camp was salty.

We stayed with Mike's friends, a couple who ran a little goat farm near Bronte Lagoon. They had met a man in 1967 who was driving from Queenstown to Hobart. At one point, he said he had seen a dog-like animal with stripes. He had never heard of the Tasmanian Wolf and had no idea of the importance of his sighting. The couple's daughter had also seen a thylacine in 1985. Whilst walking home from Deloraine when she saw the creature in a meadow.

One of their friends had a strange encounter in 2006 but one not related to the Tasmanian wolf. The woman had been kayaking in a lake Mersey River. Some huge aquatic animal swam up to her kayak. Whatever it was the thing was pulling a large wake and disturbed the witness. He thought it could possibly be a giant eel.

The following day we met up with Lloyd and Maureen Poke. The couple used to have a farm in the north east of Tasmania. When they were there from the 70s to the early 90s it was still a wild place. Now the area has been swallowed up by farms that have sprung up all around. During their years on the farm they both claimed multiple sightings of the Tasmanian wolf. Lloyd had his first encounter with the creatures as a boy in 1957 near the Ouse River. He watched a family of thylacines, a mother with three cubs, walking through the scrub. He decided not to tell anyone.

In 1986 the couple were driving along a road on their property when they saw a strange beast in the road ahead. Maureen said…

"I realized it had stripes and could not be a tiger quoll. It had to be a Tasmanian tiger."

The creature was around fifty meters from the witnesses and had a striped back and stiff tail, a description that should be familiar by now!

The following year Lloyd noticed that something was taking dead wallabies that he had shot on the property. He was using them for dog foot. Intrigued he brought one thousand feet of tough cotton and tied it to a wallaby carcass. Later, when the carcass had been taken he followed the cotton like Theseus following his ball of twine in the labyrinth of the minotaur. The cotton lead him through the forest to a lair under an old tree stump. At the time, he had no camera so he set up a tape recorder. He was rewarded by recordings of crunching noises and odd screams. Lloyd suspected that a Tasmanian wolf was responsible.

As it turned out he was correct. Sometime later he saw a thylacine chase a wallaby into the scrub. He found it biting into the wallaby's chest. The predator stood up on its hind legs and gaped its formidable jaws in a classic thylacine threat display. Lloyd backed off immediately and left the animal to its meal.

Another time he was out with his dog and crawled into a wallaby trail in the bushes. He heard a snarling from further along the trail and retreated.

Lloyd's closest encounter happened in 1990 after he had put in a new fence. He saw a Tasmanian wolf walking along the fence and managed to corner it. He tried to catch it by grabbing it.

"I grabbed at its neck thinking that there would be loose skin but it was tight and muscular. I couldn't hold onto it and it struggled free."

Turning around it kicked out scratching Lloyd's arms then leapt over the fence and disappeared. The couple left the farm in the mid-1990s.

Soon I was back on the plane en-route back to England. Already we have another expedition planned for December of 2017.

Each time I have returned to Tasmania I have had my conviction that the thylacine is still alive and well re-enforced. I have not revealed everything that happened on this last expedition. Suffice to say I have seen some convincing evidence about which I have been asked to keep quiet for the time being. I feel it is now just a matter of time before the Tasmanian wolf is officially removed from the extinct list.

Ape Canyon Feature Film Seeks Financing

My feature film will begin shooting in the next few months. It's about Bigfoot. Because of course it is.

And it needs your help.

Ape Canyon is a film born of years spent deep in the glorious weeds of cryptozoology—a journey that began in a first-grade classroom with a book about the Loch Ness Monster, and a teacher willing to say, when asked if Nessie is real, a simple but definitive yes. From there followed years of books and fascination, carrying on through elementary, middle, and high school—and then into Oberlin College, where I had the opportunity to teach what turned out to be the world's first college course on cryptozoology anywhere in the world in over a decade.

I was actually one of the top cryptozoology stories of 2004. Imagine that: http://www.lorencoleman.com/top_cryptozoology_2004.html

Now, years later, my first feature film is heading into production: a cryptozoological dramedy I wrote called *Ape Canyon*. The story of a cryptozoology buff in the midst of a quarter-life crisis was a passion project for years, carrying on in the background as other screenplays earned screenwriting awards and film options, and as my debut novel The Listeners came and went. It was as an inaugural fellow of the Johns Hopkins University/Saul Zaentz Innovation Fund that I decided to pursue independent film in earnest, and it's through my connections there that I joined forces with Josh Land and Victor Fink.

Josh and Victor are the director and director of photography of critically acclaimed, Indiecapitol Award-winning microbudget film *Lotus Eyes*.

Josh and Victor took an immediate shine to the project. With their help, we've joined forces with production company Studio Unknown, a company that includes

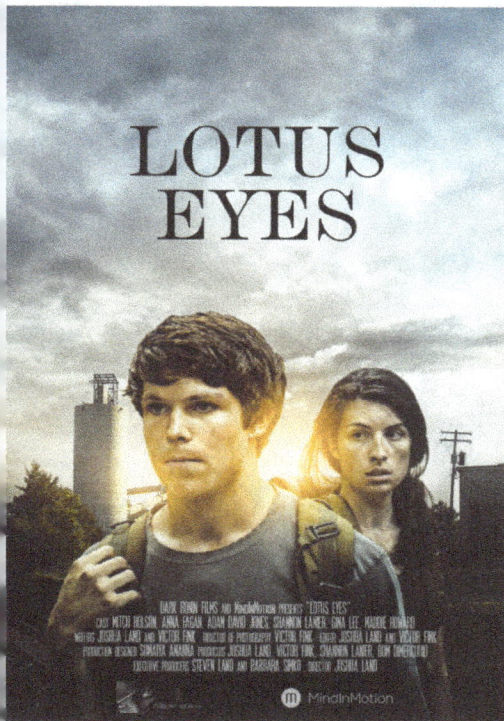

within its board of directors Blair Witch Project co-director Ed Sanchez.

We have locations in the Pacific Northwest, not too far from Ape Canyon itself. And we have a talented cast, including Jackson Trent (*Reel Iconic*, *The Cleaner*) and Anna Fagan (*Lotus Eyes*, *First & Last*) as our leads.

And we've raised $75,500 of the $85,500 we need to make the movie.

That's where you come in.

On its own $10,000 is not a small sum— but with the collective cryptozoological community, and on top of the far larger sum already raised, it is in no respect unattainable. And let me be clear here:

This is no charitable donation. Investors in *Ape Canyon* are associate producers, with shares in profits and a say in the direction of the project.

This is a story I've been working with for some time. It's one I believe in. It's one Josh and Victor believe in, and one Jackson and Anna believe in. It's one the investors of $75,500 believe in.

So the question is this: Is it a story you could believe in? Is it one you would like to see come to life?

If so, I invite you to journey with us into Ape Canyon. Bigfoot may remain elusive, but this film can and will come to pass.

To learn more about becoming involved with *Ape Canyon*, please e-mail

hdemchick@gmail.com
or
apecanyonfilm@gmail.com.

Harrison Demchick is a developmental editor, an optioned and award-winning screenwriter, an inaugural fellow of the Johns Hopkins University/Saul Zaentz Innovation Fund, and the author of 2012 literary horror novel The Listeners.

In 2004 he also became the first person to teach a college course in cryptozoology anywhere in the world in more than a decade.

Pic on left courtesy Caroline/Wikimedia Commons

DISCUSSION DOCUMENT: DUTIES FOR REGIONAL REPRESENTATIVES

The CFZ had had regional representatives for over twenty years now. Some of them have done remarkable things, some nothing at all, and some something in between. I originally intended my first wife to manage the list of regional reps, but as history shows, that never happened.

Ever since Alison and I split up I have been intending to ask someone else to take over the job, and finally a few months ago I got around to it. Ronan Coghlan has agreed to take over the onerous task, and has come up with a list of suggested roles

for regional representatives, which I post here for public discussion.

[1] In the event of a reported sighting of a mystery animal in the representative's area, all possible data should be gathered and forwarded to CFZ. Likewise, news of further developments should be sent on as they occur.

[2] Representatives should try to discover if there were any sightings or other anomalous events in their areas in the past, but should only send on stories of UFOs or ghosts if they consider them important, as otherwise their is the danger of CFZ being swamped

[3] Representatives should, if possible, look into local folklore to discover if stories of anomalous events in the area occur. Liaisons should be initiated with the Bird, Butterfly and Conservation Officer in their areas where possible. They should, in addition, try to gather an archive of Fortean zoological material from their local studies libraries.

[4] Representatives should initiate liaisons with groups dealing with anomalies and nature in the area, provided they consider them and their personnel suitable.

[5] Representatives should have the option of offering sales of books to local bookshops. However, some might find this distasteful and so this should not be regarded as an actual representatives' duty.

Letters

The editor and his compadres welcome letters for publication on all subjects covered by this magazine. However, we would like to stress that neither this magazine, or the CFZ are responsible for opinions expressed, which are purely those of the letter writer.

FROM THE HORN OF AFRICA

Hello Mr. Downes

I hope you and yours are doing well.

I was just curious if you had heard from your members on the photos I had sent you? I was going to send the video of it also, but my laptop has entered into a new dimension of its own. Whether it is a virus or it has become self aware and chooses not cooperate, I guess I will never know.

So for now I just use my tablet. I know that Africa and the surrounding coast hold many unique wonders. I often thought if marine biologists would spend 6 months just going to the pier early every morning. They would see many things. The fishermen and poachers would bring their catch their to sell. I remember hammerheads and various other sharks

flopping about on the deck. Not to mention the hundreds of other species that were brought in with their nets.

I was able on several occasions to observe them on their skiffs. They would have several car batteries in the middle of the boat. They would connect old florescent lights to them and then hang the lights above the water several feet. This brought in the bugs, then the smaller fish, and so on. Spearing or netting anything that came along. Although they were aware of what was illegal to catch and what wasn't. Daily survival and poverty will always out weigh what some scientist in San Francisco has to say. It is a very harsh reality that is hard for people to understand. I think until someone lives there for extended period of time they will never understand. I was looking over your website more today. I am hoping one day to travel to England and attend one of your conference's. I find the subject very interesting. That modern man has yet to discover new things in his own world.

If I can assist you in anyway please do not hesitate to ask.

My regards to you and yours,
Thank you for taking time to look this over.
Name withheld by Editorial Decision

EDITORIAL REPLY: The current thinking from all the people that we have consulted is that it is the remains of a pilot whale. Compare the teeth in the lower jaw of the picture on the previous page with the teeth in this museum photograph by Wolfgang Sauber.

A MUIRHEAD MISCELLANY

Dear readers

Here is a summary of recent stories I have found in the Geneaology Bank 1690-2016 newspaper archive:

This first one is from an American newspaper about an one-eyed monster but sadly I lost the date and name of newspaper. This is a partial extract.

ONE EYED MONSTER

"Clyde Ruoff has had a lot of trouble trying to prove he saw a one-eyed monster. One after-noon down the river, Clyde reports,he looked down by his feet and there, in about a foot of water, lay something with four legs and a long tail. Much to Clyde`s amazement, he could see that the strange reptile had only one eye right in the center of its head!

Well, we`d like to believe Clyde. He did say that he tried to catch the critter so he could bring it back as proof of his story. Clyde knows the rules of the club - and he must pay a forfeit for not being able to prove his TALL TALE. (Unless one of you other fishermen could be able to capture the one-eyed river monster and bring it in for Clyde`s proof.")

NEW RABBIT
(image below is a CFZ reconstruction)

Fort-Worth Star-Telegram (Texas) Nov 26th 1906

Quanah Texas

"A new species of rabbit has just been discovered in this portion of Texas. G.L.Lambert of this county captured yesterday what he thought was a common cottontail rabbit, but which turned out a complete surprise. The animal resembles its namesake in every way except that it has a small horn growing from the center of its head. This horn is about an inch and a half long and very pointed. This is the second horned rabbit captured in this section, one having been caught in Oklahoma a few years ago that had three well developed horns. The animal in question has attracted a great deal of attention and many theories as to its origin. "

ODDLY COLOURED SQUIRRELS

Harrisburg Patriot (Pennsylvania) May 19th 1896

"Some oddly colored squirrels are said to have been taken near Belleville, Ontario. A black squirrel with numerous white spots was killed by Hull Austin, and another man got a fox colored black squirrel. The queerest two were black squirrels, one with a red tail and the other with a big white spot on the breast and one on the back. A cream colored black squirrel and a "snow white" one, both rufous brown on the underparts, were killed. Such groups of odd animals are often noted in certain neighborhoods. In some places freak robins will be seen every year; in another it is oddly colored quail.

HUMAN HEAD WITH BODY OF A GATER MAN FINDS STRANGE CREATURE IN A RIVER.
Boston Journal (Massachusetts) October 1st 1909

"Bath, man finds strange creature in a river. Michael Welch picked up in the Kennebec River something this morning something which they term a mermaid. Whether it is a real mermaid or what it is nobody seems to know. They do know, however, that it is something they never saw before.

It is of a dark brown color with a small head, apparently human, about the size of a good-sized fist, with a pipe-stem neck and well developed chest also arms while the rest of the body is of alligator formation. It is the strangest exhibition ever shown in the city and scores of people flocked to the store of Charles T.Jackson during the forenoon to see it. There is no life in the object and Mr Welch and his companion at first paid no attention to it. Later they went to it as they rowed across the river and discovering that it was a freak of some sort, brought it ashore."

BOLAM BEAST REDUX

Dear Jon,

Guess what?

14 Years after a spate of Spectral Big Foot sightings in Northumbrian country park of Bolam Lake. The sightings have been immortalised in a wood carving at the main entrance. Way back in 2003 I was lucky enough to be involved (in a small way) with the CFZ investigation which resulted in a sighting of the thing that became as The Beast of Bolam. Seeing the carvings today brought it all back.

Davey Curtis

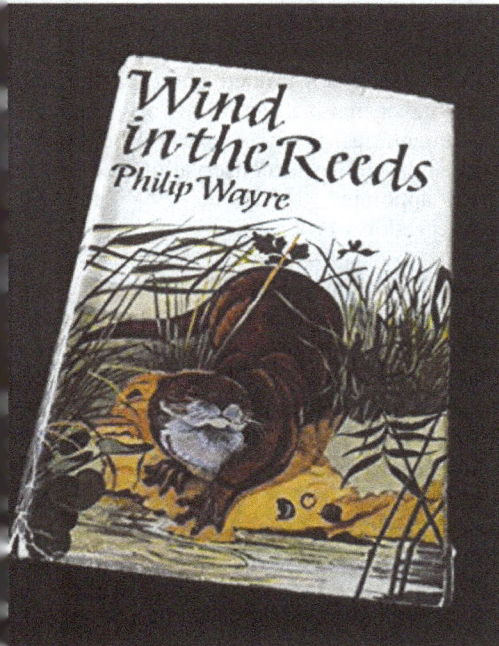

Hardcover: 255 pages
Publisher: Collins; 1st ed. edition
(1965)
Language: English
ASIN: B0000CMPXP

When I write reviews for various publications they are usually, but not always, of books that have been fairly recently published, or which at least are fairly easily obtainable and currently available. The subject of this review is none of these. In fact, belay that. It is fairly easily available; second hand through Amazon, with prices starting at £0.01. But it was published in 1965, over half a century ago, in an unimaginably different world. It was a world where Britain still had the remains of an empire, where we were still surfing the post war economic boom, where Sir Winston Churchill had only just died, and where the year had begun with a Beatles Christmas Show featuring both Rolf Harris and Jimmy Savile, and where the world (at least if you were British and relatively well off) seemed to be a much more peaceful and logical place than it does today.

And why am I reviewing this book today?

It is fairly simple. My old friend Richard Muirhead, whom I have known longer than anyone else in the world who is not a blood relative, recently gave me a copy, and reading it threw up a number of interesting aspects, which - I believe - are worth discussing in a wider forum.

I first heard about Philip Wayre in 1967, when my Godmother gave me something called 'The Zoo Annual', or something like that: a book which contained an article about Wayre's Zoo in Norfolk which was unique in that it featured only European animals. It was a time when - largely due to the influence of my hero Gerald Durrell - British zoos were heading away from the old concept of being places of entertainment, to becoming places where serious scientific and conservation research could be carried out.

I knew that Wayre had been instrumental in

PHILIP WAYRE
WILDLIFE TRUST

the founding of The Pheasant Trust and The Otter Trust, but I knew very little about him. So, after reading his book, but before writing this review, I checked out the biographical details at:
www.philipwayrewildlifetrust.co.uk,

which I make no apologies for having lifted in its entirety:

"Philip Wayre (1926-2014) was interested in wildlife from a boy when he would study rats under the joists at his prep school to watching and photographing the geese on the mudflats of East Anglia as soon as he was old enough to drive.

After service in the Royal Navy in the Second World War he settled in Norfolk to try his hand at farming, something he admitted he was never good at. His passion for wildlife remained. In the late 1950's he had a slot on Anglia Television showing animals he kept and tamed at his home. This led to Philip making programmes about natural history including many for the Survival series on Anglia Television.

His collection of animals grew and in 1961 he turned his farm into the Norfolk Wildlife Park, the first of its kind in Britain. He became well known for his breeding successes of endangered species which he released back to their countries of origin. Concerned at the decline of the otter he co-

founded the Otter Trust in 1971, a charity which pioneered the captive breeding of otters for release into the wild and has been credited with saving the otter from extinction in much of England.

Philip was a self-trained photographer and film-maker making natural history films and documentaries. He wrote several books and served on numerous conservation bodies. He was appointed MBE in 1994 for his conservation work".

Now, the thing that I find particularly interesting about this book, is that - unlike the present generation of conservationists - and indeed any wildlife professionals for the past twenty or thirty years, Wayre (like Durrell, Gavin Maxwell and numerous others) were from what was once called the Gentry, and were largely unqualified for the roles which they decided to take on through life.

These days one finds the idea of unqualified, and unlicensed amateurs running something like a zoo as unbelievable a concept as someone who drives a car without the benefit of a driving license, and with a snootfull of Bolivian marching powder. But it is undeniable that many of these men and women were true pioneers, who - despite being wracked with personal problems - achieved remarkable things.

age, and I find the way that Wayre writes, totally acceptingly, of - in particular - otter hunting, particularly disturbing. But Wayre was born nearly four decades before me, and four decades is a very long time.

I know that young people of my acquaintance who are interested in the natural sciences, are shocked and disturbed to learn that when I was a child, and even in my early twenties I collected butterflies, and so I truly shouldn't be surprised at the fact that someone born forty years before me had a different moral and cultural *omphalos* to the one that I have developed.

Gerald Durrell was an alcoholic, and Gavin Maxwell was wracked with guilt over his homosexuality (which only became legal during the last two years of his life). And I have a sneaking suspicion that there was something seriously affecting Wayre's life. Because, although I do not know anything for sure, there is a haunted quality to his writing which seems to hint that there is *something* that he was not telling us. *Something* that by the standards of the society of his time, he did not want to admit in living black and white.

Like the other zoological and natural history pioneers that I have mentioned above, Wayre came from a class to whom "huntin', shootin', and fishin'" were a standard part of their way of life. But so did I, and I wholeheartedly rejected all of these pursuits, finding them abhorrent from a very early

Because 1965 was a very different time and place to the Britain in which I, and many of the people reading this now, live in, during the second decade of the 21st century. And I think that this is the point that I wanted to make. We find a world where amateur conservationists could buy a baby sun bear without a license from a London pet shop, and co-exist with people that decent folk nowadays consider to be animal abusers totally incomprehensible.

But no doubt my two little granddaughters will find aspects of the way that I have lived my life totally incomprehensible, and quite possibly upsetting. So my message to them, and to you, is quite simply:

Context matters darlings, context matters.

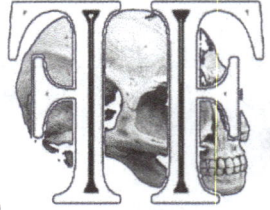

THE WORLD'S WEIRDEST PUBLISHING GROUP

We publish a lot of books. Indeed, I think that we could quite easily claim to be the world's foremost publishers of books about Fortean Zoology and allied disciplines, and our Fortean Words imprint is doing a great job in producing books on other non-zoological esoterica. However, I feel that it would be unethical to review our own titles. So here, to end this edition of *Animals & Men*, is a brief look at the books we have put out so far this year.

Paperback: 216 pages
Publisher: cfz (13 Dec. 2016)
Language: English
ISBN-10: 1909488488
ISBN-13: 978-1909488489

Richard Muirhead is one of the longest standing members of the Centre for Fortean Zoology (CFZ).

When the CFZ started a daily blog in 2009, Muirhead was one of the first contributors, and over the intervening years has contributed hundreds of fascinating articles. A trained librarian, he has what Charles Fort would have called a 'wild talent'; he has a remarkable attitude for unearthing arcane data from obscure archives. CFZ director, Jonathan Downes, has known him for nearly half a century, and considers him to be the best researcher he has ever known. Read this, the first selection of his blog postings to be anthologised and we expect that you will agree with him.

ON VAMPIRES

RONALD MURPHY JR

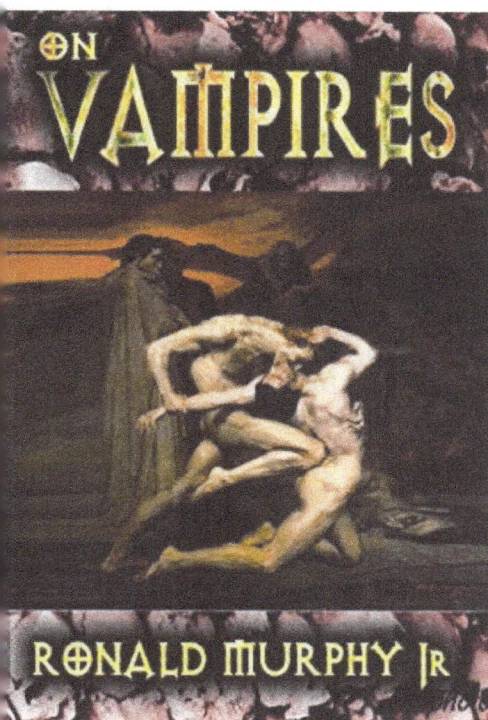

Paperback: 154 pages
Publisher: Fortean Words
(24 Feb. 2017)
Language: English
ISBN-10: 1909488518
ISBN-13: 978-1909488519

Join cryptozoologist and folklorist Ronald Murphy as he journeys throughout history in his quest to uncover the impetus for the archetype of the vampire.

Beginning at the lair of cannibals at the dawn of human history, explore the images and evolving ideas of the vampire, tracing these concepts up to the information age. Keep a stake close by as you uncover the world of the vampire.

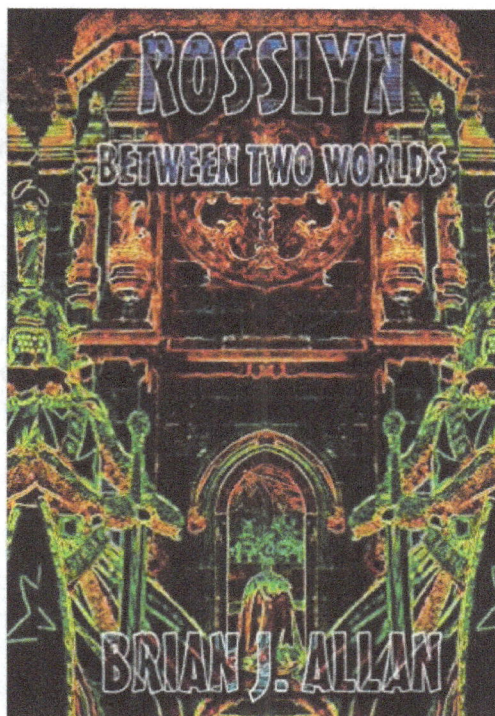

ROSSLYN BETWEEN TWO WORLDS

BRIAN J. ALLAN

Paperback: 142 pages
Publisher: Fortean Words (27 Feb. 2017)
Language: English
ISBN-10: 190948850X
ISBN-13: 978-1909488502

Over the years there have been several books written about Rosslyn Chapel, the most recent seem determined to undermine any possibility that this very special building is anything more than a stone church, albeit highly detailed and elaborate, and nothing more. This work confronts this rationalist stance head on and presents startling new evidence suggesting just the opposite. From the outset it should be clear that two quite distinct histories are applied to Rosslyn Chapel. One is the official version insisting that the Chapel is nothing more than a place of simple piety and worship, while the other is a much more esoteric shadow interpretation with immense implications, the histories are not interchangeable and there are very few points of convergence.

WEIRD WEEKEND DENMARK JUNE 3.-4. 2017
August Krogh Institute, University of Copenhagen
Universitetsparken 13, DK-2100 Copenhagen

PROGRAMME:

Lectures:
- Lars Thomas: Introduction to Cryptozoology and the Weird Weekend
- Jonathan Downes: Some classic and not so classic beings of cryptozoology
- Shoshannah McCarthy: The reality of the Chupacabra
- Ronan Coghlan: Celtic mythologi and strange Irish beasts
- Professor Tom Gilbert: DNA and Cryptozoology
- Professor Mikael Rothstein: Why do we believe in strange things
- Lars Thomas: Mystery beings of Denmark
- Ole Henningsen: The making (and faking) of UFO photos
- Inger Winkelmann: The secrets of the giant squid
- Torsten Schlichtskrull: The Bestiarium of Conrad Gesner

And all for just £15 a day.

There will also be a ghost walk, photographic and arts exhibitions, how to draw your own monster – workshop, demonstrations of natural history field work and possibly guided tours of the natural history museum in Copenhagen (the price of the guided tours are not included in the overall ticket price).

For further information about the event, the lectures and how to pay: larsthomas1960@gmail.com

Animals & Men

The Journal of the Centre for Fortean Zoology

Issue 61

THE GIANT TAILED TOAD OF DEVIL'S ARSE

Identifying felid prints; The sad story of the Blue Monkey; Hoaxes (three different ones); Warning signs on Wikipedia; A new twist on o.o.p butterflies; news, reviews and more...

Contents

Typeset by Jonathan Downes,
Cover and Layout by SPiderKaT for CFZ Communications
Using Microsoft Word 2000, Microsoft Publisher 2000, Adobe Photoshop CS.
First published in Great Britain by CFZ Press

CFZ Press, Myrtle Cottage, Woolsery, Bideford, North Devon, EX39 5QR

ISBN: 978-1-909488-31-1

Faculty of the Centre for Fortean Zoology

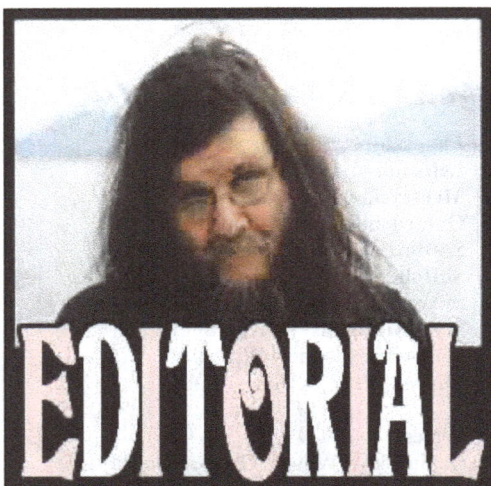
EDITORIAL

Dear Friends,

As we mark our Silver Jubilee, welcome to Issue 61 of the world's premiere cryptozoological magazine. We are, as the Chinese proverb goes, living in interesting times, and I very much doubt whether there is anybody reading this who has not been touched, somehow, by recent events on the world stage.

I always envisaged that the Centre for Fortean Zoology, the organisation which I formed so many years ago, would be a broad church, with room for people with all sorts of beliefs, and from all sorts of intellectual and philosophical backgrounds, and I have no intention of changing that. However, various things have happened recently that I have found disturbing, and I feel that I have to nail my particular colours to the mast in this very special year for the CFZ.

A few months ago, I was approached by an American TV production company who asked me whether I was interested in fronting a major series which was looking at lake and sea monsters around the world from a paranormal perspective. Now, as I have made perfectly clear over the years, although I take a very broad, and holistic view of the various "things" (if I may do an Ivan T) that we investigate, I dislike the words 'Paranormal' and 'Supernatural', because I believe that the *real* phenomena which are so often lumped together under these appellations are both NORMAL and NATURAL, but are merely the result of laws of nature that we do not yet understand.

But I am not stupid. Despite my declining health, the idea of fronting a US TV show with a substantial salary, appeals quite a lot. So, I agreed to speak to them, and they immediately asked me to sign a non-disclosure agreement; something that I always find irritating, as it implies that my word as a gentleman isn't good enough. But I did sign it, hoping that I would be rewarded with big bucks. And, as far as I am concerned, what I am writing here is in no way a contravention of that agreement. And if it is, I don't really care.

They asked me how I would approach such a series, and I unwittingly shot

The Great Days of Zoology are not done!

myself in the foot straight away, because I said that I thought that this was all about humankind's relationship to water, and that it should start small - citing the affair some years ago when snakeheads got loose in Maryland waters - and working up to Loch Ness.

There was a stunned silence.

"We are not interested in science," he said, and went on to say how he had envisaged the series being me rushing about in a helicopter talking to psychics and mediums, and people who believed that lake monsters were malevolent space aliens. They would let religion in; they wanted to feature members of a church who - being Young Earth Creationists - believed that the existence of these creatures was proof of their ideas. Indeed, they wanted the Creationist angle in. But they didn't want science.

This time it was me that produced the stunned silence.

"What's wrong? Don't you believe in God?" he asked.

"It's actually none of your business," I replied, "but as it happens, I do. But the Earth is 4.54 billion years old. Science proves that….."

"Pah. Science.." he snorted, and I realised that for the second or third time in my life, my career as an American TV presenter was disappearing down the pan. The conversation terminated soon after, and I have - not surprisingly - heard nothing from him since.

The social changes and technological advances that have taken place since the beginning of this century have transformed the way that a large proportion of our species behaves and thinks. In particular, the advance of social media has meant that everyone now has a voice, and can put anything they want into the public domain, mattering not whether it is a massive intellectual advance for humanity or a load of tosh. Humans being humans, it is usually the latter.

Like all other young hippies, I embraced the concepts of Marshall McLuhan's Global Village, but now it has become a reality, it is much like any other village. It may *look* picturesque in a picture postcard sort of way, but in the older and prettier houses the roofs leak, there is an ever-present band of yobbos by the bus stop, there is a small estate on the outskirts of the village where Social Services have rehomed a bunch of problem families, and the Squire and his family buggered off years ago, and the Manor House has become a nightclub where there are drunken fights every Friday night.

And, back in the real world, a belief in superstitious nonsense is rife, science is distrusted by far more people than I would ever have thought possible twenty-five years ago, and the world is generally a worse place to be if you are a

scientific freethinker than it was twenty-five years ago when I started the CFZ.

So, I find myself, once again, in the peculiar position of having to lay out our stall, twenty five years after we first did so.

The CFZ is - as I have said - a broad church, and I have always said that there are not, and never will be, any thought police intent on crushing incidences of CFZ Thought Crime.

The only things that are FORBIDDEN are animal abuse, cultural, human and racial/religious prejudice. But I and the rest of the CFZ Management would like to stress that we are, anti-bloodsports, and vehement opponents of Young Earth Creationism, and we believe that - here in the Anthropocene - the changes in the

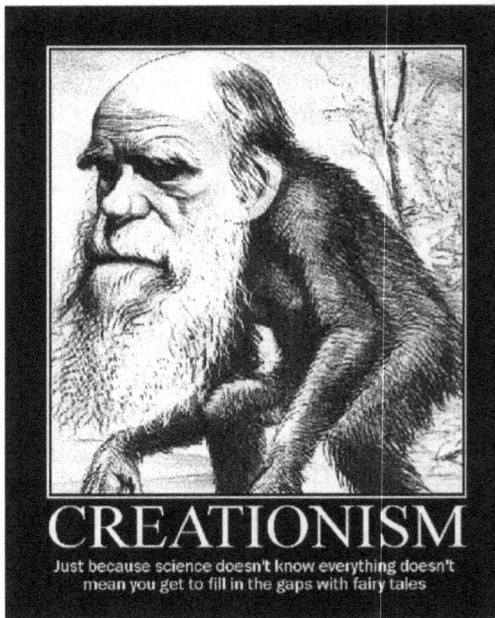

CREATIONISM
Just because science doesn't know everything doesn't mean you get to fill in the gaps with fairy tales

climate which are self-evident are at least partly caused by our own species.

How does this affect the CFZ?

When presented with blatantly racist material we will not, and never will, publish it.

Some years ago, we were sent a quite well written article claiming that British Big Cats had been smuggled in by illegal immigrants from the Middle East determined to destabilise the British farming industry and impose Sharia Law. Not only did it make very little sense, but it was obviously designed to foment racial conflict, and we returned it to the author with a polite, but curt, note explaining why we would not print it.

Although I have already stated my views

on the matter, if we were to be sent a well-written and reasoned article which had either Young Earth Creationism or Climate Change Denial as the main crux of its argument, then we would probably print it, but - as always - we would allow rebuttals and criticisms to be published alongside it.

There are those who claim that we are entering a new Dark Age and although I do not necessarily agree, I do understand where they are coming from. And in times like these it is more important than ever to proceed in a diligent and steadfast manner. And so, as we enter our second quarter century, we must shoulder our burdens and move forward.

Onwards and upwards,

Jon Downes

A LEGAL MATTER

Newsfile

You Wascally Wabbit

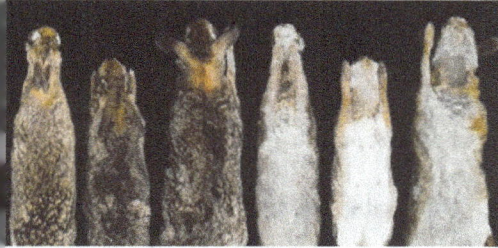

"At a time when species are going extinct every day, it's encouraging to know we can add one back on the list of survivors," said PSU biology professor Luis Ruedas, announcing the discovery of a new species of cottontail rabbit - the Suriname lowland forest cottontail (*Sylvilagus parentum*) - from South America. His findings were published on May 17 in the *Journal of Mammalogy*.

Ruedas made his discovery after studying rabbit specimens at the Naturalis museum in Leiden, in The Netherlands. The specimens, collected in 1983 from the small country of Suriname (formerly a Dutch colony until 1975, and still with significant links to the Netherlands) on South America's northeast coast, were labelled as South American cottontails.

Ruedas studied the anatomy of the museum specimens and determined they were larger and shaped differently than other rabbits throughout South America - so much so that they deserved to be classified as a distinct species.

The creature will be only the third new rabbit species named in South America since the start of the modern classification system 260 years ago.

SOURCE: https://academic.oup.com/jmammal/article-abstract/doi/10.1093/jmammal/gyx048/3828752/A-new-species-of-cottontail-rabbit-Lagomorpha

Horns of a dilemma

The Albany adder (*Bitis cornuta albanica*) is one of the rarest snakes in the world; a viper subspecies of *Bitis cornuta* or a separate species alternatively identified as *Bitis albanica*.

Its range is restricted to eastern and southern Cape Province in South Africa. Like all vipers, it is venomous. It has a brilliantly patterned body and pointy eyebrows. The extremely rare reptile hadn't been seen in almost a decade, and

scientists feared it was extinct—until now.

An expedition set out last November to find the long-lost snake, and after a week of scouring bushes, lifting up rocks, and cautiously peeking into holes, team member Michael Adams spotted a six-inch-long female slithering across the road. What's even more amazing is the team found four live animals—only 12 individuals have been recorded since the species was identified in 1937. (The scientists did find a fifth snake that had been killed by a vehicle.)

"I certainly think it's among the most threatened globally," says Bryan Maritz, a regional coordinator for the International Union for Conservation of Nature's Viper Specialist Group who wasn't part of the recent expedition.

SOURCE: http://news.nationalgeographic.com/2017/05/albany-adder-venomous-snake-extinct/

Crafty Climbing Crab

As a boy in Hong Kong I was always fascinated by the mangrove crabs which climbed surprisingly high up the trees along the shoreline of the top of Tai Tam Bay. I am particularly excited, therefore, to be able to announce the discovery of a new endemic species of micro-crab from the former British colony.

The Mangrove Ecology and Evolution Lab, led by Dr Stefano Cannicci at the

A

B

Pseudosesarma patshuni, was described in 1975.

SOURCE: https://www.sciencedaily.com/releases/2017/04/170411104534.htm

A Hell of a Snake

Swire Institute of Marine Sciences (SWIMS) and School of Biological Sciences, the University of Hong Kong (HKU), has recently discovered, described and named a new species of mangrove-climbing micro-crab from Hong Kong, and published the description in *ZooKeys*, a peer-reviewed and open access international journal dedicated to animal taxonomy.

The new species has been given the scientific name of *Haberma tingkok*, since all the specimens found at present were spotted at a height of approximately 1.5 to 1.8 metres above chart datum, walking along the branches of the mangroves of the Ting Kok area. The crabs are small, less than a centimetre long, predominantly dark brown, with a squarish carapace, very long legs and orange claws. It represents the second endemic mangrove crab species described in Hong Kong. The previous one,

A new paper, published in the open access journal *ZooKeys*, which describes a total of three new species of Atractus groundsnake from Ecuador. One of them is the Cerberus groundsnake, (*Atractus Cerberus*). It is predominantly brown in colour with faint black longitudinal bands, and measures about 21-31 cm in length. The biologists justify the curious name of this species with the peculiar location where they spotted the first known specimen. Found at the gates of the newly formed "Refinería del Pacífico," a massive industrial oil-processing plant, the authors were quick to recall the multi-headed monstrous dog Cerberus, known to be guarding the gates of the underworld, according to Greek mythology.

SOURCE: https://www.sciencedaily.com/releases/2017/03/170322103717.htm

Hyalinobatrachium yaku is a newly discovered species of frog in the Centrolenidae family. It is found in the Pastaza, Orellana and Napo Provinces of Ecuador.

One of the remarkable characteristics of this species is that their belly and some internal organs including the heart are transparent. The glassfrogs are generally small, ranging from 0.8 to 3 inches (2-7.5 cm) in length.

It was described, by five researchers namely Juan Manuel Guayasamin, Diego F. Cisneros-Heredia, Ross J. Maynard, Ryan L. Lynch, Jaime Culebras and Paul S. Hamilton first published about the finding on May 12, 2017, in the journal *ZooKeys*.

SOURCE: https://en.wikipedia.org/wiki/Hyalinobatrachium_yaku

Another Compendium of Batrachia

A new species of Asian mountain toad belonging to the genus Ophryophryne has been discovered in the the Truong Son or Annamite mountains of Vietnam, an area of high diversity for the group. The toad, one of the smallest species of horned mountain toads ever described to science, was given the name *Ophryophryne elfina*, which roughly translates to "elfish eyebrow toad" — and the researchers who made the discovery say that there is evidence to

Another Compendium of Batrachia

suggest that the species could already be considered endangered.

A team of Russian and Vietnamese researchers described *Ophryophryne elfina,* the Elfin mountain toad, in the journal *ZooKeys.*

The species name elfina, of course, refers to elves — small, magical forest creatures found in German and Celtic folklore. The new toads have horn-like projections above their eyes and are as diminutive as their namesake: at around three centimeters in length, they are the smallest known species of the genus Ophryophryne.

SOURCE: https://news.mongabay.com/2017/06/new-elfin-mountain-toad-discovered-in-annamite-mountains-of-vietnam/

New species are rare enough, but a new genus?

A group of researchers in Indonesia has created a new genus of arboreal toad to fit what they say are two new species, according to a recent article in the journal *Herpetologica.*

The authors propose that the newly described toads, *Sigalegalephrynus minangkabauensis* and *Sigalegalephrynus mandailinguensis*, be classified under the genus *Sigalegalephrynus*, after finding indications that the two "form a distinct lineage" among Southeast Asian members of the true toad family *Bufonidae*.

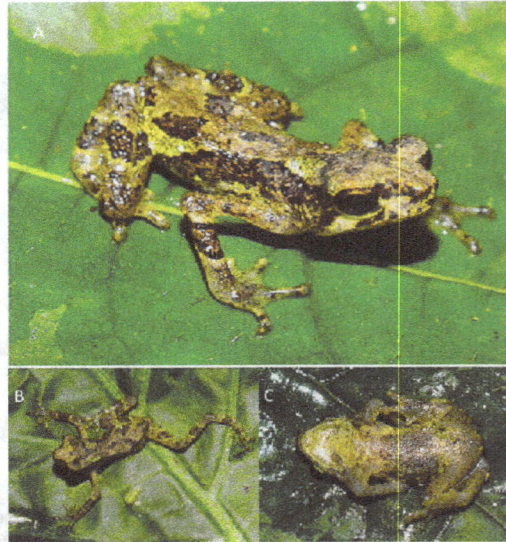

The researchers — from the Indonesian Institute of Sciences (LIPI), the University of Texas, Broward College, Brawijaya University and Hamburg University — came across the species while inventorying the reptiles and amphibians of Sumatra's highlands in 2013-2014.

SOURCE: https://news.mongabay.com/2017/04/new-genus-created-for-arboreal-toads-in-indonesia/

Another Compendium of Batrachia

Sneaky Squirrel

There are now three species of flying squirrel in North America, and it turns out that the newest member of the family has actually been gliding amongst the treetops of the U.S. West Coast, sometimes right alongside its closest relatives, this whole time.

Researchers described the new species in a study published in the *Journal of Mammalogy* in May. *Glaucomys oregonensis*, or Humboldt's flying squirrel, can be found all along the Pacific Coast, from southern British Columbia all the way down to the mountains of southern California. The squirrel had previously been classified as northern flying squirrels (*G. sabrinus*) due to their similar appearance. A genetic analysis revealed the coastal populations belong to a distinct species all their own. "For 200 years we thought we had only had one species of flying squirrel in the Northwest — until we looked at the nuclear genome, in addition to mitochondrial DNA, for the first time," Jim Kenagy, professor emeritus of biology at the University of Washington (UW) and a study co-author, said in a statement.

SOURCE: https://news.mongabay.com/2017/06/dna-analysis-reveals-a-third-species-of-flying-squirrel-in-north-america/

Flying forest frugivore

Whilst most of their relatives are carnivorous some leaf-nosed bats pursue a vegetarian lifestyle, dining on fruit, nectar, and pollen. One of these fruit-eating – or "frugivorous" – groups is the Sturnia genus, which holds the record for the most species. Sturnia, commonly called yellow-shouldered bats, received yet another species recently, after researchers from institutions in Venezuela and Ecuador looked more closely at the genetics and bones of bats in northern Colombia and Venezuela.

The researchers discovered that two groups of bats that had been classified as a certain species were in fact members of another species unknown to science. They describe this new species, which they named *Sturnia adrianae*, in a study published recently in the journal *Zootaxa*. The species had previously been lumped into *Sturnia ludovici*. The new species has already been divided into two subspecies: *S. a. adrianae* and S. a. caripana. The former is larger and was found in western and north-central Venezuela, as well as across the border in Colombia; *S. a. caripana* is smaller both in size and distribution, with its range restricted to a small area of northeastern Venezuela.

S. a. caripana is restricted to the Turimiquire Massiff, an isolated mountain range to the east of the broader range of *S. a. adrianae*. The area is subject to slash-and-burn clearing, a technique in which farmers burn vegetation to convert an area for crop production. So the newly described bats are already under threat.

SOURCE: https://news.mongabay.com/2017/04/new-leaf-nosed-bat-uncovered-amidst-burning-habitat-in-venezuela/

Liminal Loach

This pink, scaleless and possibly blind cave loach is the first ever instance of a fish found living in a European cave, and those behind the discovery said it is most northerly species of cave fish ever found. According to a report on the discovery in the journal *Current Biology*, the cave loach likely split off from surface-dwelling loaches at within the last 20,000 years.

The fish was first sighted in a difficult-to-reach section of a subterranean water system in Southern Germany by amateur diver Joachim Kreiselmaier, who snapped a photo of the fish and passed it along to Jasminca Behrmann-Godel, an expert in fish evolution at the University of Konstanz in Germany. "When I saw the photo I wasn't sure it was really something special," Behrmann-Godel told *BBC News*. "Then he brought me a live specimen and that was like the bang. That was the moment we realized that this was something really new!" "No more than 30 divers have ever reached the place where the fish have been found," Kreiselmaier noted.

"Due to the usually bad visibility, strong current, cold temperature, and a labyrinth at the entrance, most divers do not come back again for diving."

SOURCE: http://www.redorbit.com/news/science/1113417847/first-ever-european-cave-fish-discovered-by-amateur-diver/#zs3jHA15Hd3QvLQc.99

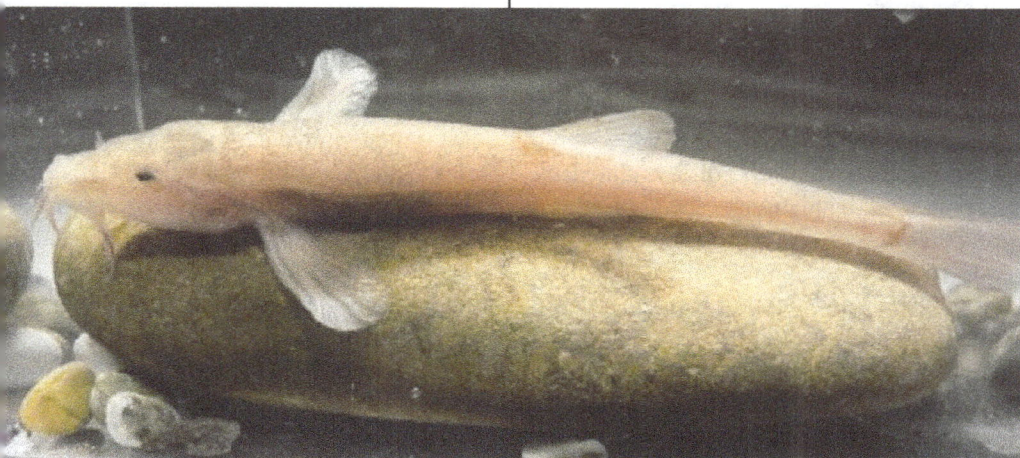

Several Ounces

A recent research paper in the *Journal of Heredity* reveals that there are three sub-species of snow leopard. Until now, researchers had assumed this species, *Panthera uncia*, was monotypic.

Studying snow leopard scat from wildlife trails and marking sites revealed three primary genetic clusters, differentiated by geographical location: the Northern group, *Panthera uncia irbis*, found in the Altai region, the Central group, *Panthera uncia uncioides*, found in the core Himalaya and Tibetan Plateau, and the Western group, *Panthera uncia uncia*, found in the Tian Shan, Pamir, and trans-Himalaya regions. This is the first range-wide genetic analysis of wild snow leopard populations.

The snow leopard is considered the world's most elusive large big cat and inhabits a vast area of around 1.6 million km2 across 12 countries in Asia. It is a high-altitude specialist that primarily occupies mountains above 3,000m in elevation, a habitat characterized by low oxygen levels, low productivity, temperature extremes, aridity, and harsh climactic conditions. The snow leopard is the largest carnivore in its high-altitude habitat in many areas and is under substantial threat throughout its range.

SOURCE: https://www.sciencedaily.com/releases/2017/05/170511141941.htm

Magnificent monitor

Scientists have recently found and re-described a monitor lizard species from the island of New Ireland in northern Papua New Guinea. It is the only large-growing animal endemic to the island that has survived until modern times. The lizard, *Varanus douarrha*, was already discovered in the early 19th century, but the type specimen never reached the museum where it was destined as it appears to have been lost in a shipwreck.

The discovery is particularly interesting as most of the endemic species to New Ireland disappeared thousands of years ago as humans colonized the island.

The monitor was discovered during fieldwork by Valter Weijola from the Biodiversity Unit of the University of Turku, Finland, who spent several months surveying the monitor lizards of the Bismarck Islands.

It can grow to over 1.3 metres in length and, according to current information, it is the only surviving large species endemic to the island. Based on bone discoveries, scientists now know that at least a large rat species and several flightless birds have lived in the area.

"In that way it can be considered a relic of the historically richer fauna that inhabited the Pacific islands. These medium-sized Pacific monitors are clearly much better at co-existing with humans than many of the birds and mammals have been," says Weijola.

SOURCE: https://www.sciencedaily.com/releases/2017/05/170502095832.htm

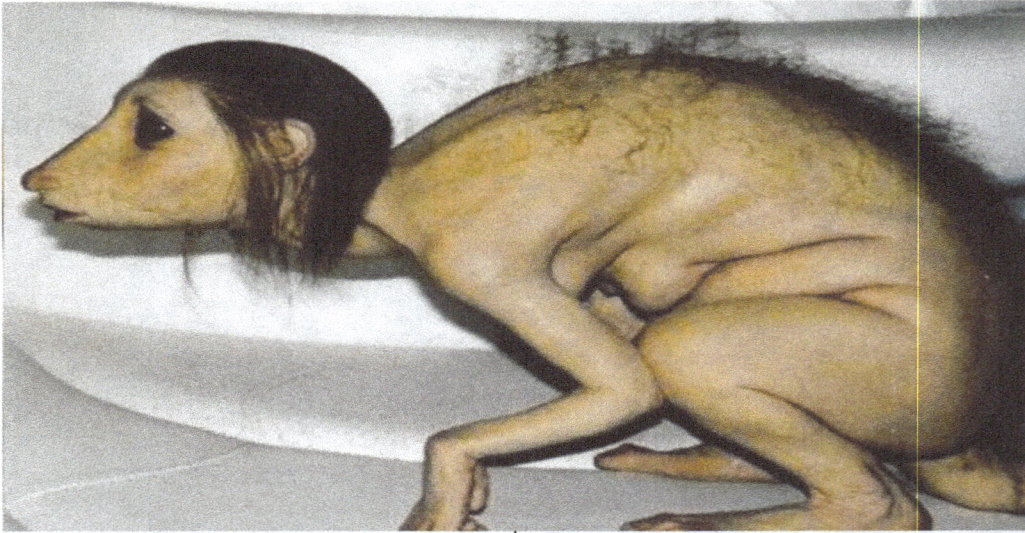

Chupacabras

Mystery Image

I like this picture which is oddly cute in a Jim Hensonesque way, but it is hard to imagine that anyone is going to take it seriously. Peculiarly, it has turned up a lot recently, mostly accompanying general articles about the history of the chupacabras phenomenon, such as this one:

https://evonews.com/info/2017/may/17/video-chupacabra-still-terrifies-the-world/

I would be interested in finding out where this image came from, so if anyone reading this can enlighten me, I should be very interested to find out more.

Honduran Horror

In May we received reports of allegedly vampiric attacks on livestock from the village of Choluma in Honduras.

The story runs as follows: "In the past two months the beast has been responsible for the deaths of at least 35 animals, all drained of blood and brutally slaughtered.

Omar Martinez told *La Tribuna* about his nephew who cares for a heard of sheep woke up at 12:45 in the morning when he heard a noise. Investigating further he saw a dark shape surround the sheep who fell dead shortly after. Paralyzed with fear the boy watched as the shape moved on to kill two more goats, eventually leaving".

The original report gives only a few more details (naming the nephew as Nely David Martinez, for example, but seems to

corroborate the account that we were sent from *Creepy Times*.

SOURCES:
http://www.latribuna.hn/2017/05/06/chupacabras-desata-furia-una-aldea-choloma/
http://creepytimes.com/creep-o-pedia/cryptozoology/el-chupacabras-en-choloma-no-bueno/

Cordoban Chicanery

The next story comes from Valle de Punilla, in Córdoba, Argentina, also in May. The National Food Safety and Quality Service (SENASA) claims to have finally solved the mystery effecting farmers where several cows and horses were found dead, apparently after being attacked by some sort of animal predator. A good start one might think. The report reads: "The strange bites on the animals' bodies, mostly on the neck and the underside, were immediately attributed to the chupacabras, a creature from Latin American folklore that, according to legend, drinks the blood of livestock, particularly goats".

The news item then appends a picture that appears to be of a horse with a bloody great hole in its rib cage, and what appears to be a

cored rectum. Again impressive. That is until you read the final dénouement:

"However, experts from the SENASA confirmed yesterday that the deaths are caused by an outbreak of a type of rabies transmitted by common vampire bats, located in San Marcos Sierras."

Has nobody told them that vampire bats are tiny creatures? Or has an enterprising editor just put the story together with pictures that have been pinched - almost at random - from the Internet? The horse picture appears to be from at least 2012 when it was used to illustrate a classic horse mutilation case in Argentina. It can be found in a number of online locations including:

http:// inexplicata.blogspot.co.uk/2012/07/ argentina-vision-ovni-investigates.html

Inexplicata is a journal which I admire greatly, and have never even suspected them of any impropriety.

Politicians talking nonsense? Or journalists writing nonsense? You pays your money and you takes your choice. However, what is this peculiar picture that looks like it is vaguely of an aye aye? Once 'auto levels' are applied using Adobe Photoshop. it looks more like a fat moggy than anything else. And to confound things further, a brief use of Google Image Search does tend to imply that this picture, at least, dates from May 2017, and was first used to accompany various reports of this story. What it is, however, is an entirely different kettle of fish.

SOURCE: http://www.thebubble.com/ senasa-finally-solves-chupacabra-mystery-in-cordoba/

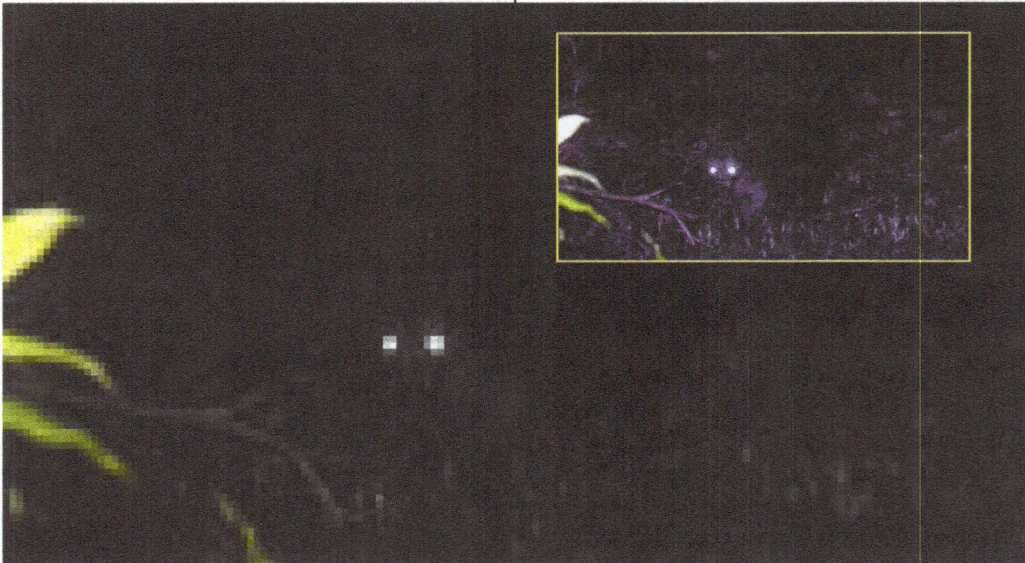

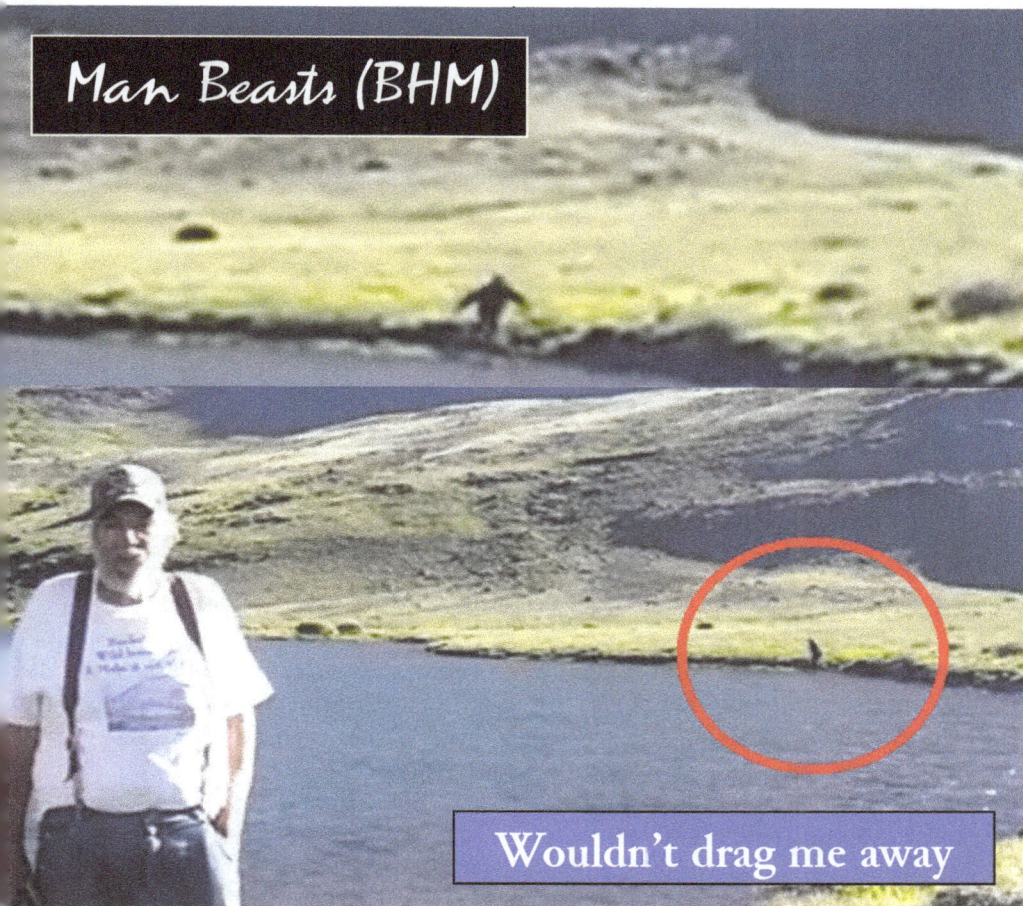

Man Beasts (BHM)

Wouldn't drag me away

We will take any excuse for a Gram Parsons inspired headline here at *Animals & Men*.

In mid April a sequence of video footage was unleashed upon an unsuspecting world. The footage is said to have been taken during a family trip to Oregon state in the US, and shows a man talking to camera as he is next to Wildhorse Lake in the Steens Mountains. And the *Daily Express* then claims: "But, way behind on the other side of the lake, the unmistakable alleged form of Bigfoot can be seen". How something can be both alleged and unmistakeable we are not too sure, but it does appear to show a large hominiform object moving along in the middle distance.

SOURCE: http://www.express.co.uk/news/weird/792609/Clearest-BIGFOOT-caught-camera-Wildhorse-Lake-Oregon

Double Dutch

I have always been interested in British and European accounts of BHM phenomena, although - except in the far north of Finland, and the Caucasus - I think it is highly unlikely that these are bona fide cryptids.

Instead I believe that they are a composite collection of various types of zooform phenomenon, hoaxes and misunderstandings.

In April a piece of video from Holland, filmed in a Dutch national park by two youngsters named Lucas and Jeroen grabbed the public attention. *The Sun*, which - sadly - is not the most reputable British newspaper, takes up the story.

"The lads were wandering round Veluwezoom National Park when the noticed the humanoid figure peering at them from behind a tree in the distance.

The pair can be heard swearing with incredulity when they see the creature lurking in the forest."

Somne commentators have claimed that the 'creature' is wearing "white sneakers" but neither my eyesight or my concentration span is good enough to confirm this..

SOURCE: https://www.thesun.co.uk/news/3426053/bizarre-footage-shows-dutch-youngsters-freaking-out-while-filming-a-bigfoot-peering-out-from-behind-a-tree/

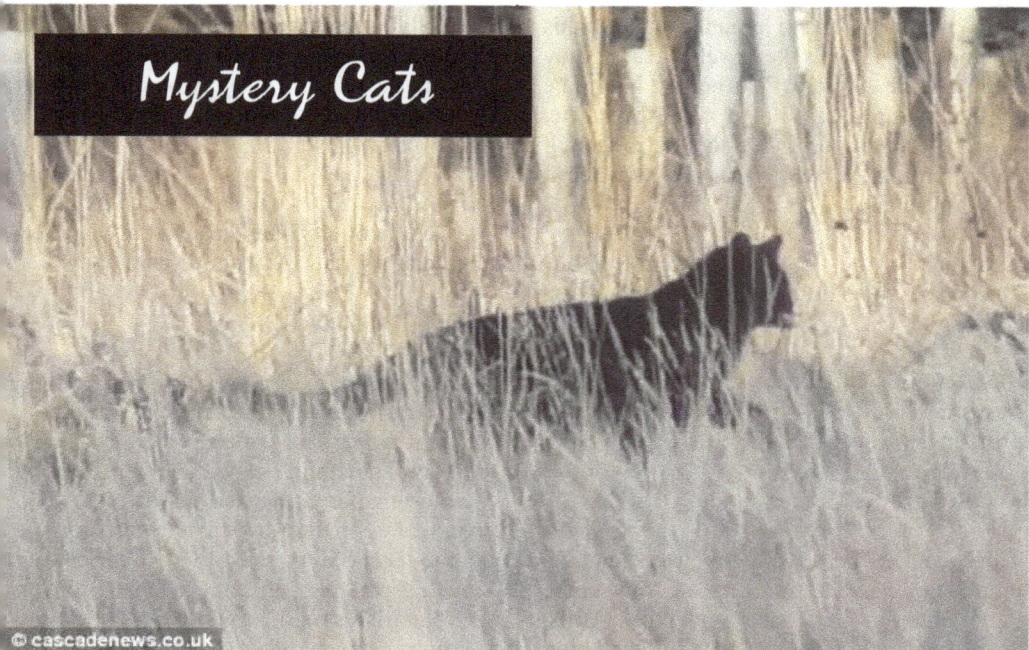

Mystery Cats

© cascadenews.co.uk

- INVERNESS:

A postman has spotted a 'big cat' roaming the fields on the outskirts of Inverness. Paul Dawson took a picture of the black cat with his 300mm long-lens camera at the weekend in a field close to Milton of Leys in the Scottish Highlands. He says he was standing around 600yards away from the animal with a long black tail that appeared much larger than a domestic cat when he pictured it.

However Doug Richardson from the Highland Wildlife Park believes the black cat was a domestic tabby. He said: "The posture, tail position and apparent thickness, size of the animal and its apparent shoulder height in relation to the grass are consistent with this. It would be amusing to remind readers of Doug Richardson's history with British big cats, especially the skull found on Bodmin Moor twenty two years ago, but it would be unkind.

SOURCES:

http://www.dailymail.co.uk/news/article-4355904/Postman-says-photographed-big-cat-Scotland.html

http://www.highland-news.co.uk/News/Big-cat-sighting-a-domestic-tabby-says-wildlife-expert-31032017.htm

- WARWICKSHIRE:

This picture did the rounds again this summer, but it was taken over a year ago by Philip White, 39 in a field behind his home in the picturesque village of Great Alne, Warks. The dad-of-one said: "I've seen it before but never had my camera so when it appeared again I just started filming".

And no to the journalists at the *Daily Telegraph* who said that it looks like a lynx. It doesn't!

SOURCE: http://www.telegraph.co.uk/news/2016/11/10/lynx-like-cat-filmed-prowling-warwickshire-field/

- SOMERSET

This "panther" pictured below was seen by a mum enjoying some sunshine at Crowcombe Park Gate, in Somerset's Quantock Hills. The 32-year-old, who was with her daughter, 14, said sunlight was glinting off the animal's fur and it prowled through a patch of long grass. The woman, who did not wish to be named, said she immediately took a photo of it.

SOURCE: http://www.dailystar.co.uk/news/latest-news/606083/panther-big-cat-someset-quantock-sightings-uk-jaguar-wild

- HERTFORDSHIRE

Sharon Smith, 53, was stuck in Hatfield Road traffic on Friday evening at about 6.30pm, when she glanced out the window and saw something moving in the field by the thoroughfare. She said it was big, sleek, cream, and moving towards the car. Doing a double-take, Sharon tried to alert her partner - but he was driving and concentrating on the road. She said: "I could see its back and it was creeping along and I thought, 'that

doesn't look like anything I have seen before'.

SOURCE: http://www.hertsad.co.uk/news/ big-cat-spotted-creeping-through-st-albans-field-1-5029961

Valerie Rodrigues was on her way back from Redbourn at about 8.30pm on June 6 when she spotted an animal dart out into the road in front of her car.

At first she dismissed it as a fox - it was a "gingery, sandy colour" she said, a similar size, and trotting - but as the car got closer it started to bound away.

She is now "absolutely sure" it was feline: "It was definitely a cat in the way it moved, in its well-defined muscles in its legs.

"Cats don't run like dogs or foxes, they are more arched in their backs. I don't doubt it was the cat.

SOURCE: http://www.hertsad.co.uk/news/ muscular-big-cat-spotted-bounding-across-road-near-st-albans-1-5064568

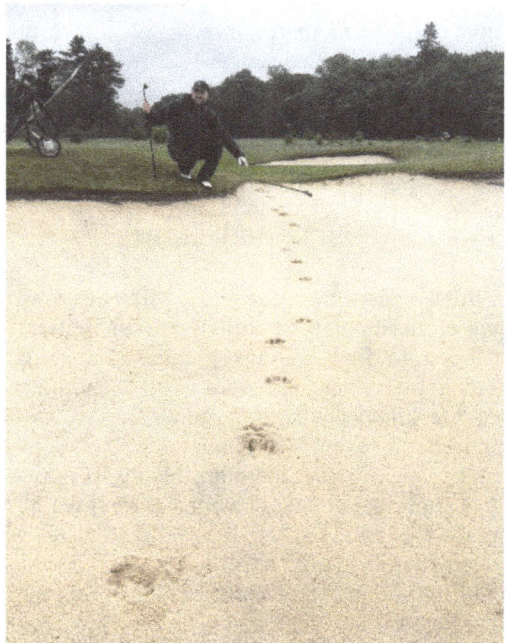

Wheathampstead local Stuart Braggs, 50, was playing golf on the Aldwickbury course when he noticed something "bizarre" by one of the bunkers near the clubhouse. Animal footprints almost as large as his palm lead in long strides across the pit - preserved in the wet sand after a rainy morning.

He said: "We took photos and carried on playing golf, but I was then aware that some large cat had been walking there, which puts you a little bit on edge." "There's no way it was anything else but a cat, it's definitely something reasonably large."

SOURCE: http://www.hertsad.co.uk/news/ photos-of-large-pawprints-are-new-evidence-of-st-albans-big-cat-1-5061442

Alistair Divall, of Old London Road, saw what he believed to be a panther while he was out jogging, at around 9.30am. He said: "It was about 10 or 15 feet in front of me, moving slowly, and then it turned and looked at me briefly. "It disappeared completely and there were people down at the bottom with their dogs."

SOURCE: http://www.hertsad.co.uk/news/jogger-reports-seeing-panther-on-alban-way-today-1-5055062

- WARWICKSHIRE REDUX

Those readers from the West Midlands who were disappointed earlier in this news section to find that the pictures touted as being from this year were actually nothing of the kind can breathe anew. For we do, indeed, have big cat pictures from Warks. this issue. At least there are blobby pictures that look like a newly hatched axolotl which may be of a big cat.

Chilling footage shows the beast stalking the harvested field next to University Hospital Coventry. Stunned Matt, 36, was waiting for partner Hayley to recover from giving birth when he spotted the creature outside the window on Wednesday. He told *Sun Online:* "I was just glancing out and saw this thing moving across the field. It was absolutely huge. I've been around dogs and seen plenty of foxes in my time and this creature just wasn't moving like that."

We would be interested to know quite what makes this amorphous blob so chilling. But, hey. Who are we to cast aspersions?

SOURCE: https://www.thesun.co.uk/news/3883001/new-dad-stunned-after-capturing-what-he-believes-is-the-notorious-wildcat-of-warwickshire-on-camera/

Aquatic Monsters

Less than Ness

Even by the standards of *The Daily Star* the following headline is bad: "Is that the Loch Ness? Viewers terrified by hellish footage of underwater monster". Presumably they mean "Is that the Loch Ness Monster?" but in any case the answer is a resounding "NO, OF COURSE IT ISN'T YOU SEMI LITERATE SCUM BADGERS!" Loch Ness is, after all, in Scotland whereas this story takes place in Florida.

The bloke seen at the beginning of the story is grinning like a monkey, and is obviously far from being "horrified", and the "monster" is obviously a shoal of fish of some kind (or possibly a school of some small cetacean). However the "Daily Fail" claimed that it was either immigrants, benefit claimants or single parent families come to steal good white people's jobs and fish food.

SOURCE: http://www.dailystar.co.uk/news/latest-news/624027/Loch-Ness-video-proof-viewers-horrified-water-beast-caught-on-camera

Chinese rocks

One of the least well-attested monster reports of recent years comes from China's Zhelin Reservoir near Jiujiang. A video which was apparently taken by swimmers, shows what many have claimed "looks like a giant sea creature

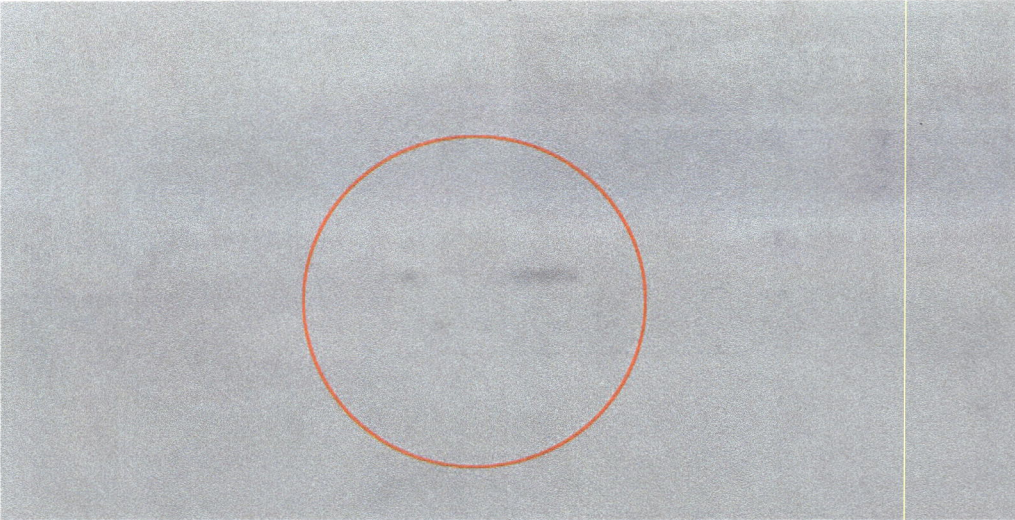

or reservoir dragon" partially emerging from the waters. The video was uploaded on May 25 and the identity of the witnesses and actual date of the sighting are not given. According to one media report, the location appears to be the shore of the Mount Lu Xihai Resort in Jiujiang City in East China's Jiangxi Province. The video is blurry but the audio clearly provides the sounds (in Chinese) of excited serpent spotters.

The report implies that they're saying the creature is 16 to 19 feet long. No other sightings of this Zhelin Reservoir monster seem to have appeared in the media, so there's not much to go one in determining what exactly the video is showing.

SOURCE: http:// mysteriousuniverse.org/2017/05/ reservoir-monster-dogs-swimmers-in-chinese-reservoir/

Ness than Perfect

There have been a number of sightings of anomalous objects which could be interpreted as being of the Loch Ness Monster in recent months. The first took place at the end of April, and is taken from Gary Campbell's site.

28 April - A Ms Cairney form Dunbartonshire was at the loch with three of her friends when they saw a 20m long series of waves move at about 5mph along the surface 500m out. It came out of nowhere and then disappeared the same way - they have confirmed that there was no boat traffic in the area.

The sighting took place at 3pm from a layby on the A82 road between Drumnadrochit and Inverness.

SOURCE: http://www.lochnesssightings.com/index.asp?pageid=498361

Fast forward to the beginning of May when Hayley Johnson, a 28-year-old care assistant from Abbey Hey, Manchester, saw a strange and dark shape at dusk in the loch's Urquhart Bay. She is quoted as saying:

"I am a really sceptical person. I have never even been to Loch Ness before, but I decided to come up for the Bank Holiday.

I had stayed in a backpackers' hotel and on my last night decided to go for a walk through the woods and ended up on the banks of the loch. It was lovely and at dusk.

Then about half a mile away I saw this dark shape sticking up – like a neck. I thought at first it was a tree, but it was very strange. I took a picture. It was there for a couple of seconds, but when I looked back it was gone. I was shocked.

I was really excited about Nessie as a child but to be honest I thought Nessie had probably died in the 1930s. I didn't think she was alive any more. I know now that she is very much alive. I'm just so excited – it's unbelievable what's happened."

SOURCE: https://www.thesun.co.uk/news/3461385/loch-ness-monster-sighting-first-time-in-eight-months/

Then six days later came this:

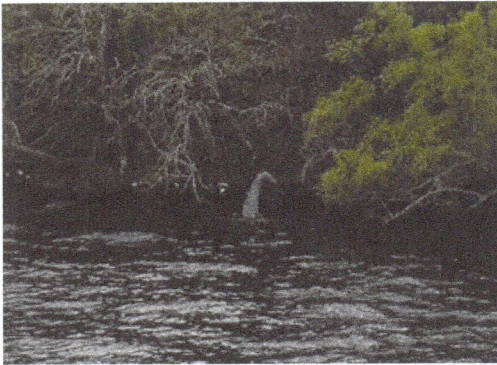

The Loch Ness Mystery Blog notes: "Apparently this was photographed in the vicinity on the 7th May (six days later). It has a very papier-mâché look to it and one wonders if an Adrian Shine type experiment is going on?" I have not been able to find any further information about this picture.

SOURCE: http://lochnessmystery.blogspot.co.uk/2017/05/is-nessie-back.html

BUT. And, yes it is a big but (and I am not going to make any jokes about big butts, except for the ones we use to collect rainwater) then, within a day or so came this. To quote from *The Metro*:

"....he (or she) glides through the water barely causing a ripple as a boat floats by. There's someone on the boat that goes by, but they don't seem all that phased by the appearance of the legendary creature. The commentary on the video, shot on May 7, suggests that there were many witnesses to the sighting, convincing us further that the Loch Ness monster totally exists and is a real thing. The footage was uploaded to YouTube by 'Sweetie' who asked in his description: 'What the hell was it? Did anyone else see it?'"

SOURCE: http://metro.co.uk/2017/05/11/loch-ness-monster-may-not-be-dead-after-all-as-new-footage-emerges-6630313/

And finally, a few days before we went to press, Australian tourists Peter Jackson and Phillippa Wearne were driving alongside the loch in the Highlands when they saw something big and fast moving through the water. The couple quickly snapped the image of "Nessie" on a smartphone and showed it to a local skipper who told them he'd "not seen anything like it".

Phillippa, 60, a retired lawyer, from Sydney, said: "I really was just stunned and I thought, 'what is it?' "It was pretty big even from 150 yards or more offshore. I didn't know what to think. We took photos and showed them to people at a B&B and then showed them to people on a cruise."

Gary Campbell is quoted as saying ""With regard to Peter's pictures, as with pretty much all Nessie photos, they are just that little bit indistinct. "However, the report that he has submitted gives much more detail on the distances and time frames and from this, there is really no clear explanation as to what the family caught on camera."".

SOURCE: https://www.thesun.co.uk/news/3893114/loch-ness-monster-caught-on-camera-by-stunned-holidaymakers-as-the-mythical-beast-is-photographed-for-the-second-time-this-year/

A Merman I should turn to be

I am indebted to those jolly nice people at Creepy Times for this peculiar story which apparently broke in April. A Youtube Channel called APEX TV, claims that this video shows a mermaid like creature, reportedly caught on camera in a freshwater lake in the north-eastern United States.

I cannot improve on what the *Creepy Times* reporter says:

"The video itself has been make its making through the usual circles and at first glance is quite odd.

First, the clip that we're allowed to see is quite short, nothing is available of the before and after of the big reveal of the tail. Second, the water is quite excited around where the tail shows up, what's this from? Third, the tail itself seems hinged on two leg-like bones under the skin, either pointing towards a fake costume or some sort of biology we don't understand.

The concept of a freshwater mermaid is incredibly appealing, with so many massive water ways yet to be explored, who knows what lurks in their deeps. Is this the sign of something like this?

I don't know. Given how little we know of the location, the footage conditions, the person who recorded it, it's all very difficult to tell."

SOURCE: http://creepytimes.com/ creep-o-pedia/cryptozoology/video-possible-freshwater-mermaid-sighting/

Teratology

A Sense of Porpoise

Two-headed conjoined porpoises were recently hauled up by a fishing boat in the North Sea, not far off the coast of the Netherlands. The bizarre-looking creatures were already dead, and the fishermen, who are reported to have "feared trouble from the authorities", took photographs of the Siamese cetaceans, and then tossed the duo back overboard.

"It's not clear exactly why the porpoises died, but the double-headed creatures likely could not swim", said Erwin Kompanje, a researcher at the Natural History Museum and the Erasmus MC University Medical Center in Rotterdam, who wrote up the case in the June 7 issue of the journal Deinsea. Another cause may have been their fused hearts, which may not have worked adequately, Kompanje said. Conjoined cetaceans are very rare, and this may be the first case of conjoined harbour porpoises on record.

SOURCE: https://www.livescience.com/59495-two-headed-conjoined-porpoises-discovered.html

Newsfile Xtra

ENCYCLOPAEDIC KNOWLEDGE

Wikipedia is both a boon and a bane to the 21st century journalist and academic alike.

The fact that it can be edited by anybody is, in theory, magnificently Egalitarian, but in practice mean that any self-opinionated idiot can write whatever they want; something that has irritated me on many occasions. I have had people claim, on my page, that I am married to an American called Lisa. I have changed this on a number of occasions, pointing out that I only know one American called Lisa, and she is married to a friend of mine. Whilst I did -indeed- have a girlfriend

WIKIPEDIA
The Free Encyclopedia

called Lisa for a few months about 20yrs ago, I have been married to a lovely lady called Corinna for the past 10yrs. At one time, my Wikipedia page also claimed that I was a communist; something else that is completely untrue.

But these sorts of complaints are sadly common these days, and can be seen across the internet from academia at the top to 4chan at the other end of the spectrum, so I'm saying nothing new.

One of the things that I, however, have big problems with, is the way that the references system works. It was only whilst working on our edition of George Eberhart book on cryptozoology that I found out quite how many of the websites sited in the original were now non-existent, or -at the best- at different addresses. This took a long time to rectify but -as far as I can see- there is nobody doing this at Wikipedia, so an irritatingly high number of their references are useless.

However, the thing that I want to focus upon now is one specific page: List of mammals of Great Britain, which is so inaccurate, it beggars belief.

Just a few examples:

- Lagomorpha – it only sites there being one species in Britain; the mountain hare. What about rabbits? What about the brown hare?
- Rodentia – it has red squirrels, but

what about grey squirrels? I am sure that somebody will say "but they were introduced". Well, so are three other rodents in this list.

But it is the ungulates that have caused me to come the closest to apoplexy. It is a bit different to know where to start. But I will try:

- I truly find the inclusion of various ponies highly questionable. They are not wild, but their owners allow them to roam free for much of the year. It is the same with sheep on Dartmoor, Exmoor and other places.
- It completely ignores fallow deer, but includes moose because a private nature reserve in Scotland released a pair in 2008.
- The Atlantic grey whale is listed even though it was extinct in European waters over 1500 years ago, but a whole list of other mammals that were also extant within that time frame have been ignored.

The references are cursory and mostly irrelevant and all in all, the page is a disgrace.

As an experiment, I am not going to even attempt to edit the page myself, rather deciding to watch and wait and see what happens.

Watch this space.

Postscript: And so, as I said I would, I sat and waited. Three weeks later and some of it has been fixed, and it is considerably less horrible than it was. However, it still includes the nonsense about the pair of moose released into a private nature reserve

in Scotland in 2008.

As far as I'm concerned that is, by no sensible set of criteria, any sort of introduction, any more than are the lions of Longleat, or the pelicans in St James' Park.

Unfortunately, because the species is now included on Wikipedia as being part of the British mammal list, it provides grist to the mill for people that believe that Scottish lake monsters are actually misidentified sightings of this species.

This may well be the case in Scandinavia, or in North America where there is no doubt that this species exists.

But they do not exist in the United Kingdom, and, to the best of my knowledge, have not done so in historic times, although there are fossil remains of related species in various collections.

I am going to continue to keep an eye on this page and see what develops.

Newsfile Xtra

HOAXED 1: Round Rock Bigfoot

In recent months, those people from the totally rational end of Cryptozoological commentary have, very politely, raised their eyebrows at the fact that I publish quite a few links to the more (how do I say it?) gullible end of the bigfoot hunting community. .

I do this totally intentionally for although I believe that there may well be a bona fide zoological mystery surrounding bigfoot sightings in certain part of the united states, and that there almost certainly is a zoological mystery to solve in western China and central Asia, the vast majority of what is written about bigfoot is complete hokum.

But, from a sociological point of view, and even a sociopathological point of view, the activities of, and belief systems generated by, the bigfoot research community as a whole are totally fascinating.

Because of a very successful campaign of media manipulation, bigfoot has the same sort of position in the global public consciousness as did Gray Aliens 20yrs ago

ROUND ROCK PARKS SURVEILLANCE

FREEMAN PARK
06-10-2017

in the weight of Whitley Strieber's massively popular book *Communion* in the 1990s.

I commented, 20 years ago, after my first visit to Puerto Rico that the 'alien ethos' had imprinted itself on the collective consciousness of the island, and whilst 1997 (as the 50th anniversary of the Roswell incident) grabbed the public imagination all across the developed world, nowhere did it reach quite so feverish proportions as it did on the island.

Now, bigfoot, or at least the concept of bigfoot, seems to be doing much the same. Something that I find particularly interesting is the way that the bigfoot ethos has recently been taken up by the advertising industry, and several high profile hoaxes which have been reported in recent months have turned out to be nearly parts of an ingenious publicity campaign. But until now, these

publicity campaigns have been purely on a commercial basis. Now, all this has changed.

The intriguing series of 'Bigfoot' prints found in the parks of a Texas city have been revealed to be a clever promotional campaign. The Parks and Recreation Department for the city of Round Rock, Texas generated numerous headlines this week after they posted photos of 'mysterious' giant footprints allegedly spotted by their surveillance system. However, a number of elements to the story suggested that claims of a Bigfoot roaming the area were less than genuine.

Now a report from the TV station KXAN confirms suspicions that the entire affair is part of a plan by the city to try and get local children to make use of the area's parks this summer. According to the segment, park rangers will be planting 'Bigfoot clues' throughout the city over the next few weeks and any kids who find these 'Sasquatch signs' will receive prizes.

HOAXED 2: Photoshopped thylacine

When Richard was in Tasmania earlier in the year, he and Mike Williams were shown some footage which purported to be of the supposedly extinct thylacine, which many cryptozoologists – including me and Richard – believe is one of the mystery animals most certain to exist.

There have been a number of alleged thylacine videos in recent months but, as I

pointed out in an article for #12 of the CFZ Newsletter that came out on the 1st of October last year, both of the videos are fundamentally flawed. The first of them shows a creature filmed in a surprisingly built up suburb in the Adelaide Hills. The video appears to have been edited in a peculiar fashion, and it is truly impossible to tell whether whoever did so was being deliberately misleading or was just being inept. The second video, whilst superficially far more impressive, on examination shows nothing more exciting than a young fox with mange.

However, a still photograph which was published anonymously on YouTube and Facebook on 31st May really did set pulses racing... for 5 minutes at least. A series of articles on the excellent www.wherelightmeetsdark.com.au swiftly analysed and debunked the photograph. A reader of the website called Dean Alex identified that the thylacine in the YouTube clip was practically identical to one of the better known archive photographs of the creature, which was taken in 1911 by

somebody known as Mr Tucker. The video was published by 'Thylacine Sightings Australia' with the caption "this image was captured near Ellendale in Tasmania in 2016." It was accompanied by an audio soundtrack consisting of a digitally altered voice talking about the photograph and the property on which it had been allegedly taken.

No sooner had the photograph been debunked than the YouTube account was closed, and there is no evidence that I'm aware to say who the perpetrator was. Sad but true.

HOAXED 3: A classic case

Desmond Robert "Bill" Leak (9 January 1956 – 10 March 2017) was an Australian editorial cartoonist, caricaturist and portraitist. Brought up in Condobolin and Beacon Hill, Sydney, Leak attended the Julian Ashton Art School in the 1970s. His cartoons were first published in 1983 in *The Bulletin* and after he drew for *The Sydney Morning Herald* until 1994, when he was recruited by News Limited to contribute to *The Daily-Telegraph-Mirror* and later to *The Australian*.

Leak's editorial cartoons for *The Australian* were at the centre of several controversies. Works that received considerable media coverage include a 2006 cartoon drawn during the West Papuan refugee crisis, a series of cartoons in 2007 that featured Kevin Rudd as Tintin, a 2015 cartoon depicting starving Indian people attempting to eat solar panels and two cartoons in 2016, one an illustration of a neglectful Aboriginal father and another that depicted same-sex marriage campaigners wearing rainbow-coloured Nazi uniforms.

He was also a celebrated practical joker, but it seems that this was a genetic trait, for in an editorial to mark Leak's untimely death, a columnist known as 'Jack the Insider' wrote:

"I gather his father was a gifted prankster. He was a postmaster who travelled around south eastern Australia with his young family.

Bill was born in Adelaide but spent his early childhood in Condoblin in the central west of NSW and later Goroke in the Wimmera. Around those parts there had been several vague sightings of a creature that was said to bear some resemblance to a Tasmanian Tiger.

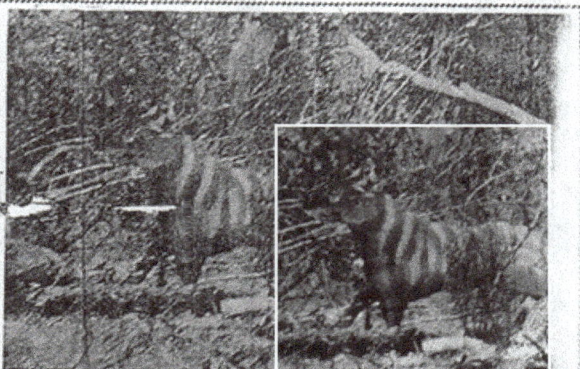

Here it is! Ozenkadnook Tiger photographed

A Melbourne woman has photographed a striped, strongly-muscled animal — believed to be the Ozenkadnook Tiger — in dense scrub near Goroke, in the Western Wimmera.

The photograph is the first ever taken of the Goroke "monster," for years the talk of district farmers and naturalists.

It was taken by Miss Rilla Martin, of Melbourne, during a holiday with her cousin, Mr Graeme Martin, stock agent, of Biely-st., Goroke.

Miss Martin saw the animal on the edge of thick scrub, 11 miles west of Goroke.

Hands shaking, she quickly raised her camera, sighted and pushed the trigger.

The *Mail-Times* photographic department enlarged the picture yesterday.

The picture shows a heavily-muscled animal with a squat, powerful head and heavy tail.

Its stripes are similar to those of the thylacine, the near extinct Tasmanian tiger.

Miss Martin took the photograph near the former home of Mr Howard Hinch.

Mr Hinch claims that "sheep" were often hauled

animals have been sighted along the eastern and southern coasts of Australia since the late eighteen hundreds.

He has compiled a book on the sightings, most of them in a scrub belt extending from southern Queensland to the south-east of South Australia.

In a note to her Goroke cousin, Miss Martin said: "This is the photograph of the thing I took that

The Wimm... Mail-Times

Bill's dad and a mate fashioned a thylacine shape out of cardboard, painted it up with the stripes roughly in the right places and armed with a box brownie, took a snap of their homemade Tassie Tiger in the scrub. The photo was passed around a small group of people as a little in joke. A bit of fun.

The world was taken in by the cardboard thylacine. Everyone knows pranks can quickly get out of hand. Somehow the grainy image of the thylacine had found its way on to the front pages of the local newspaper.

In what was a pre-internet case of going viral, the photo of the fake thylacine went from the Wimmera Mail to the pages of *The Sun* in Melbourne and from there it quickly went global. An eminent British zoologist with a list of academic credentials as long as your arm was so fascinated that he immediately jumped on a plane and flew to Australia to commence a hunt for the thylacine.

Clearly matters had taken an unexpected turn. Bill's dad pulled his young son aside and told him gravely, "You must never speak of this to anyone. Never tell a soul. Not while I'm alive."

The cardboard thylacine was stashed away in the shed and ultimately disposed of. Bill's dad and his mate got their alibis straight and if ever asked about the photo, they offered the Sergeant Schultz defence. In time the fuss died down and the Leak family moved to Sydney."

SOURCE: http://www.theaustralian.com.au/opinion/blogs/jack-the-insider-the-prank-that-took-53-years-to-debunk/news-story/e95ff01c0bdc0b023f5539aa4447769e

watcher of the skies

CORINNA DOWNES

Rare bird visitors

On 15th April, a first-winter American herring gull (*Larus smithsonianus*) was seen in the pig fields around Great Livermere, Suffolk, which is a first for the county. It is an extremely rare bird in English terms, with only 16 acceptances from eight counties (to the end of 2015). This is a is a large gull that breeds extensively across the northern areas of North America, where it is treated by the American Ornithologists' Union as a subspecies of herring gull (*L. argentatus*).

It can be found in a variety of habitats including coasts, lakes, rivers and garbage dumps. Its broad diet includes invertebrates, fish, and many other items. It usually nests near water, laying around three eggs in a scrape on the ground. Most birds winter to the south of the breeding range as far as Mexico with small numbers reaching Hawaii, Central America and the West Indies. Vagrants have reached Colombia and Venezuela and there is a report from Ecuador and another from Peru. The first European record was of a bird ringed in New Brunswick which was caught on a boat in Spanish waters in 1937 and there are have a number of additional records from Western Europe since 1990. The first British record was in 1994 in Cheshire.

SOURCE: http://www.birdguides.com

Conservation

England's first 'Swift City' takes flight

To help reverse the decline in swift (*Apus apus*) numbers and nesting sites Europe's biggest conservation charity has teamed up with nine partners to launch England's first 'Swift City' in Oxford.

Every year the swift completes a 6,000 mile migration from central and southern Africa to nest and raise their young in the UK. These iconic species land only to breed, and fly up to 500 miles per day often eating, sleeping and even mating in the air. However, with falling population numbers there are now less than 87,000 breeding pairs arriving in the UK, down from almost 150,000 (-47%) pairs just two decades ago.

Part of this decline is being linked to a reduction in potential nesting sites – as old buildings are renovated and new ones are built they often don't include nesting space. The two-year project will see the Oxford Swift City team take a closer look at the city's swift populations, their nesting sites and important foraging areas, as well as helping local communities take action to help swifts where they live.

SOURCE: https://www.rspb.org.uk/our-work/rspb-news/news/441031-englands-first-swift-city-takes-flight-

Counting albatrosses from space

Scientists have started counting individual northern royal albatrosses (*Diomedea sanfordi*) from space, the first time ever that the global population of a species has been assessed from orbit. By using the highest-resolution satellite images available, scientists from Britain and New Zealand are calculating the number of the endangered albatrosses, which breed on New Zealand's remote and virtually inaccessible Chatham Islands.

The study, published in *Ibis*, used the DigitalGlobe WorldView-3 satellite, which can observe objects as small as 30 cm across, to locate and count the birds. With a body length of over a metre, the adult albatrosses only show up as two or three pixels, but their white plumage makes them stand out against the surrounding vegetation and, for the researchers, it's a case of counting up the dots.

Results from the Chatham Islands found that satellite-based counts of Northern Royal Albatross in the 2015/2016 season were similar to ground-based counts undertaken on the Forty-Fours islands in 2009/2010, but much lower than ground-based counts undertaken on The Sisters islands in 2009/2010, which is of major conservation concern for this endangered albatross species as it may represent a major population decline in just a few years. Completing annual breeding surveys via satellite is

cheaper, less labour-intensive and requires minimal logistical effort. Additionally, satellite monitoring means no disturbance to the breeding seabirds.

SOURCE: http://www.birdguides.com/webzine/article.asp?print=1&a=6346

Project Godwit

Twenty-five rare black-tailed godwits (*Limosa limosa*) were released into their new home in the Cambridgeshire Fens in June, by conservationists from RSPB and the Wildfowl & Wetlands Trust (WWT) as part of 'Project Godwit'.

After the eggs were removed from nests and hatched in incubators, staff at WWT Welney Wetland Centre hand-reared the young birds until they were old enough to look after themselves, and it's the first time the conservation technique, known as 'headstarting', has been used in the UK.

The surrogate human 'parents' have been able to safely raise far more chicks than the godwits themselves, away from the dangers of predators and flooding, as well as – importantly - thereby prompting each pair of godwits to lay a second clutch, giving the parent birds a chance to raise a brood of their own.

Now the hand-reared birds have been released, they are expected to meet up with other black-tailed godwits hatched in the area this summer, and spend several weeks feeding on the rich wetlands before starting their migration to Spain, Portugal and West Africa.

SOURCE: http://www.cambstimes.co.uk/news/project-godwit-sees-25-rare-birds-released-into-the-cambridgeshire-fens-by-the-rspb-and-the-wildfowl-and-wetlands-trust-at-welney-1-5059257

I'iwi at Risk: A Scarlet Bird's Dangerous Migration

The native bird of Hawaii, the Hawaiian honeycreeper (*Drepanis coccinea*), or I'iwi, is classified as vulnerable on the International Union for the Conservation of Nature Red List, and environmental impacts are turning the seasonal migrations deadly, putting it at risk of extinction.

Avian malaria is thought to be the most urgent threat facing the I'iwi. It was introduced to Hawaii in around 1826, and the combination of invasive feral pigs and climate change create conditions that favour the malaria bringing mosquitoes, putting the bird at a higher risk.

Of all the honeycreepers that have been tested, I'iwi are the most vulnerable to avian malaria, and current optimistic projections hold that I'iwi will be on the verge of extinction shortly after the turn of the century.

SOURCE: https://www.islandconservation.org/iiwi-hawaii-migration/

Former drug users turn conservationists to save the Philippine Eagle

Former drug users, indigenous people and conservationists have formed an group to restore the forest where the critically endangered Philippine eagle (*Pithecophaga jefferyi*) – or Haring Ibon - lives. The aim is to plant mango, rambutan, guyabano, langka, and coffee, as well as native trees such as narra and duhat not too far away from where Philippine eagle sightings have taken place

"By planting trees," said Sam Manalastas, Community Organizer for Haribon, "they are helping not only the biodiversity of Mt. Mingan, but also their municipality in becoming resilient against climate change."

And for the conservationists, it is a small contribution to the health of Mt. Mingan, home to the King of the Birds.

SOURCE: http://www.birdlife.org/asia/ news/former-drug-users-turn-conservationists-save-philippine-eagle

45% of Arctic shorebirds are disappearing - here's the plan to save them

The Pacific Americas Shorebird Conservation Strategy aims to identify the threats and develop strategies to save the 45% of Arctic-nesting shorebirds, from across the globe, whose numbers are decreasing.

Shorebirds such as plovers, oystercatchers, sandpipers, godwits, curlews can be found along the entirety of the Pacific coast of the Western Hemisphere during some time of the year, and many species travel from Arctic breeding areas to spend their winter on the beaches and mudflats of North America, Central America and South America, where they share the environment with resident species.

The habitats they depend upon are exposed to an increasing myriad of anthropogenic threats. Within the Pacific Flyway, 11% of shorebird populations face long-term declines; none are known to be increasing.
The PASCS follows a logical sequence of setting shorebird conservation targets, identifying major threats and identifying highly effective actions to restore and maintain shorebird populations throughout the Pacific Americas Flyway, and is being led by an international group of more than 85 experts in 15 countries, including BirdLife and some of its partners.

SOURCE: http://www.rarebirdalert.co.uk/ v2/Content/Birdlife-45-per-cent-of-Arctic-shorebirds-are-disappearing.aspx? s_id=74423013

One quarter of Bangladesh is safe from recently exposed vulture-killing drug

Diclofenac, caused the most dramatic bird decline in modern history, wiping out over 99% of Asia's vultures in the 1990s. It is used by owners of livestock to alleviate pain in their animals. Unfortunately, once these animals die and are consumed by vultures, these drugs cause excruciating pain, kidney failure, and death to the birds.

All four of Asia's resident vulture species have been listed as Critically Endangered since the Diclofenac problem was exposed in the early 2000s, and through the SAVE Partnership (Saving Asia's Vultures from Extinction), BirdLife and the RSPB (BirdLife UK) have been working to ban Diclofenac in Asian countries and tackling other endangered vulture conservation issues, including creating protected "Vulture Safe Zones". However, a group of other replacement non-steroidal anti-inflammatory drugs have been exposed that pose a lethal risk to vultures, such as "ketoprofen".

Diclofenac was successfully banned in Bangladesh in 2010, and a further drug "aceclofenac" has similarly been outlawed. "Banning aceclofenac was another important step," says Chris Bowden, Programme Manager of SAVE, and RSPB. "It combated what can only be described as a 'cynical exploitation of a loophole' by drug companies, as ceclofenac is quickly converted to deadly Diclofenac in a treated animal. This has been demonstrated experimentally and published by our SAVE

research team."
SOURCE: http://www.birdlife.org/asia/
news/one-quarter-bangladesh-safe-recently-
exposed-vulture-killing-drug

Six-year-old raises money in aid of critically endangered bird

Abigail Court (6) visited the state of Victoria's Moonlit Sanctuary zoo in Pearcedale last year and was so moved by the plight of almost-extinct and critically endangered orange-bellied parrot (*Neophema chrysogaster*) that she felt she had to do something. So she raised more than AU$1,000 towards the conservation of the bird. She began a 5-cent coin drive with her local girl guide organisation which eventually raised AU$1,002.70. She also gave a speech at her school, Bayswater West Primary, and discussed the personal charity initiative with her local Salvation Army.

The orange-bellied parrot now breeds at only one site in Tasmania, having been extirpated from southern mainland Australia, where the very small population of fewer than 30 wild birds still winters. The species has been radically reduced in numbers by the spread of psittacine beak and feather disease and drought in its wintering areas. A captive population has been established as insurance against it completely dying out in the wild and to boost the remaining group. The Moonlight Sanctuary is trying to breed the parrots in captivity, with the aim of having 18 pairs this year, though the zoo only has 18 individuals at present. The money raised by Abigail will be used to pay for veterinary expenses in testing the birds before they released into the wild.

Her mother Rebecca told Melbourne's *Herald Sun* newspaper that: "She's always been an animal fanatic. She has visited a lot of zoos, but I have never seen her quite so passionate about one animal. She was fascinated by them; she didn't want them to land on her but she loved looking at them."

Well done, Abigail!

Poaching/ Illegal killing

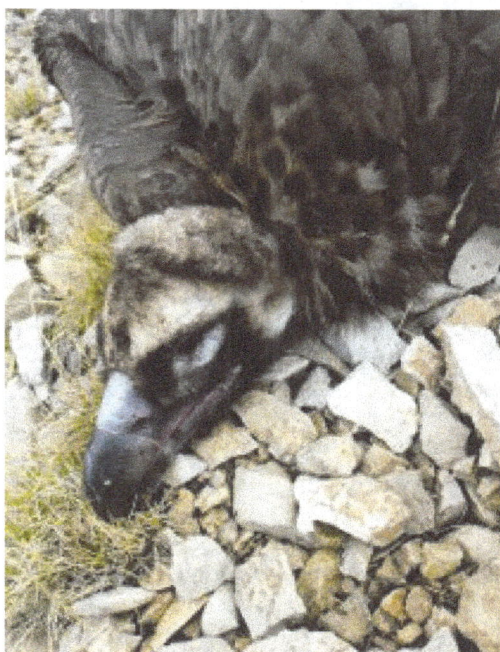

A poacher kills the last Black Vulture in Alinyà, Spain

One individual has more than likely destroyed the efforts and investment of years of work into the possibility of re-introducing the black vulture back to Alinyà, Spain. In November 2016, Trasgu, a 6-year-old male that was transferred from a rehabilitation centre to the Pyrenees in 2013 and released there as part of the project to reintroduce the species in the Pyrenees (Alinyà and Boumort), was found dead at Alinyà. The necropsy confirmed that the bird had been shot dead, in the same area that has seen two other black vultures disappear for the same reason. The news was kept secret so that an investigation could start, and was only made public recently.

The black vulture reintroduction project started in 2006 in Alinyà and Boumort - a collaboration between the MAGRAMA (Spanish Ministry of Agriculture, Food and Environment), the Foundation Catalunya–La Pedrera, the NGOs TRENCA and GREFA and the staff of the Reserves of Boumort and Muntanya d'Alinyà, and aimed to establish a new population in order to create a corridor for the gene-flow among already-existing populations in the South West of the Iberian Peninsula and the reintroduced populations in France (Cevennes and Baronnies).

SOURCE: http://www.rarebirdalert.co.uk

"Stop the Killing": Lebanese President speaks out for migratory birds

Lebanon's President, Michel Aoun, has made a heartfelt pledge to prevent the annual slaughter of the thousands of migratory birds who fly over the small Middle Eastern state twice a year; some 2.6 million birds disappear every year, shot or trapped illegally. There are at least 399 species of birds recorded in Lebanon, and they are a huge concern for local conservationists, such as those who work at the Society for the Protection of Nature in Lebanon (BirdLife Partner).

The country lies on the west side of the African Eurasian Flyway (Red Sea – Rift Valley Flyway) which is considered one of the most important flyways in the world for bird migration. The long perilous journey from Europe and Asia to Africa, via the Sinai and the Red Sea, ends here, in this small stretch of land, for millions of birds. In terms of "intensity"- birds killed per square kilometre - Lebanon ranks third, trailing only Malta and Cyprus.

But Lebanon has announced a new bird-friendly era, the announcement coming straight from the Lebanese President, Michel Aoun, with a heartfelt appeal to put the country's nature first: "It is a shame to turn Lebanon into a wasteland without plants, trees, birds and sea animals, and cutting off trees to erect buildings is a major crime" he said. "There should be a peace treaty between Man and the tree as well as Man and birds, because we continue to transgress upon them".

The issue of course is illegal hunting, rife in many areas. According to the President, "There should be a hunting season assigned from September to December, with the State exercising strictness in its execution".

SOURCE: http://www.rarebirdalert.co.uk/ v2/Content/Birdlife-Stop-the-Killing-Lebanese-President-speaks-out-for-migratory-birds.aspx?s_id=74423013

Migratory birds face extinction as key refuelling areas lost

The loss of mudflats in the Yellow Sea is the major cause of decline in migratory waders using the EEAF.

Populations of iconic birds that spend most of the year in Australia have been declining for decades, despite conservation efforts, and a new international study has pinpointed one big reason why.

"Migratory shorebird populations are plummeting in Australia," lead author Dr Colin Studds said. "However, the thing affecting their populations is actually happening thousands of kilometres away in north-east Asia."

Dr Studds, an Assistant Professor at the University of Maryland, Baltimore County, said the new study showed that a critical factor in the shorebirds' decline was how dependent they were on mudflats in the Yellow Sea, between China and South Korea, during migration. Many birds follow a migratory path from their non-breeding grounds in Australia to breeding sites in the Arctic, via rest stops in the Yellow Sea—a corridor known as the East Asian Australasian Flyway (EEAF).

"These birds may spend several weeks refueling before they continue their migration," Dr Studds said.

The researchers analysed citizen science data collected between 1993 and 2012 on 10 key species to see if a relationship emerged between reliance on the Yellow Sea as a migration stopover and rate of population decline. What they found was dramatic. The more a species relied on the Yellow Sea mudflats, the faster they were declining.

SOURCE: http://birdlife.org.au/media/loss-of-key-shorebird-refuelling-areas

Blue-throated macaw: new nesting sites

New nesting sites for the critically endangered blue-throated macaw (*Ara glaucogularis*; previously *Ara caninde*) have been discovered in the wet wilderness of the

Beni Savanna, Bolivia, in the south-western corner of the Amazon basin. The expedition was led by Bolivian NGO Asociación Armonia, the World Land Trust's partner in the country.

The birds were found close to Armonia's Barba Azul Nature Reserve, where they had been observed feeding and roosting. However, they don't breed within the protected area and their nesting sites were unknown. Reserve Manager Tjalle Boorsma led the exploration, focusing on the Motacú palm forest islands in an area where 15 birds had been found roosting north of the reserve in 2016.

SOURCE: http://www.birdguides.com/ webzine/article.asp?a=6323

Last stronghold of Scotland's iconic Capercaille at threat from huge housing development?

RSPB Scotland is seeking assurances a new housing development planned for the

cyclists.

Capercaillie will avoid areas where they might be disturbed because this causes them stress; this leads to a reduction in their available habitat which can prevent them from breeding. There are only between 1,000 and 2,000 Capercaillie left in Scotland with an estimated 85% of them being found in Strathspey.

SOURCE: https://www.rspb.org.uk/our-work/rspb-news/news/440307-rspb-scotland-raises-concerns-about-new-highland-housing-estate? utm_source=rss&utm_medium=feed&utm_campaign=News

Highlands, that would roughly double the size of Aviemore, will not harm capercaillie (*Tetrao urogallus*) in their only remaining Scottish stronghold.

Planning permission is currently being sought by 'An Camas Mòr LLP' for 1,500 houses to be built next to the Rothiemurchus Estate, in the Cairngorms National Park, that could result in an additional 3,000 people living in the area. RSPB Scotland has submitted a response to the Cairngorms National Park Authority on the planning application, raising concerns about the An Camas Mor development and asking for information about potential negative impacts on the nearby capercaillie population and how these will be mitigated.

Strathspey is the last remaining Scottish stronghold for the endangered capercaillie, a bird which is strictly protected under Scots and international law, and is one of the most iconic species in the national park. Several forests close to the proposed development are home to much of the country's surviving Capercaillie population. However these birds - which nest and often feed on the forest floor - are particularly sensitive to disturbance, especially by dogs, walkers and

Major breakthrough in attempt to unlock secrets of ultra-elusive seabird

A BirdLife Pacific team has managed to capture and tag a Beck's petrel (*Pseudobulweria becki*) – one of the world's rarest and least-known birds, and listed as critically endangered. It's hoped 'Pato' will lead us to its still-unknown breeding grounds, which will provide the essential insight to protect this rare species.

There is an estimated global population of less than 250 of this bird, was known only from two specimens observed in 1928, until it was rediscovered by Hadoram Shirihai, the

Israeli ornithologist, in a remote corner of the Bismarck Archipelago, north-east of mainland Papua New Guinea. "Ever since the rediscovery, the big question has been: where does it breed?" says Chris Gaskin, who led the BirdLife expedition to satellite tag a Beck's Petrel. "Finding these breeding sites is key to the future conservation of the petrel. But first we have to live-capture birds, and to do that we aim to capture them at sea".

To aid this BirdLife formed a partnership with the national conservation authority for Papua New Guinea (CEPA), the New Ireland Province, the Auckland Museum of New Zealand and other seabird biologists experienced in seabird capture. Together with the support of funders the Critical Ecosystem Partnership Fund, the Mohamed bin Zayed Species Conservation Fund, the Pacific Development Conservation Trust the Biodiversity Consultancy and BirdLife International a multi-disciplined eight-person team mounted a ten day expedition in April to catch and fit satellite transmitters to Becks Petrels. But with patience comes great rewards. On April 26th, a single Beck's Petrel, dubbed 'Pato' from the local name (pato lonbon – the duck of the sea) – was captured and fitted with a satellite transmitter. "Our team's reaction to our capturing the bird was classic - yells, laughing, dancing, hugs, high fives, the works!" says Gaskin.

SOURCE: http://www.rarebirdalert.co.uk/ v2/Content/Birdlife-Major-breakthrough-in-attempt-to-unlock-secrets-of-ultra-elusive-seabird.aspx?s_id=251355362

Conservation partnership launches "floating islands" in bid to save rare duck

An unprecedented partnership of organisations from industry and the conservation sector has come together in a bid to save the common scoter (*Melanitta nigra*) as a breeding bird in the Highlands of Scotland. The birds, which breed on the edges of a small number of lochs, will be helped by the creation of artificial floating islands made from redundant materials from fish farms. It is hoped that the scoters will choose to nest on the islands and this will make the nests safer from the unwelcome attention of predators and the risk of being flooded.

Source: https://www.rspb.org.uk/our-work/ rspb-news/news/441512-conservation-partnership-launches-quotfloating-islandsquot-in-bid-to-save-rare-duck

Tracking Devices Reduce Warblers' Chances of Returning from Migration

The tools ornithologists use to track the journeys of migrating birds provide invaluable insights that can help halt the declines of vulnerable species. However, a new study from *The Condor: Ornithological Applications* shows that these data come at a cost—in some cases, these tracking devices reduce the chances that the birds carrying them will ever make it back to their breeding grounds.

Journal reference: Douglas W. Raybuck, Jeffrey L. Larkin, Scott H. Stoleson, and Than J. Boves (2017) Mixed effects of geolocators on reproduction and survival of Cerulean Warblers, a canopy-dwelling, long-distance migrant. The Condor: May 2017; DOI: 10.1111/evo.13212

New and Rediscovered

Blue-winged Amazon: A new parrot species from the Yucatán Peninsula

In 2014, during a visit to a remote part of the Yucatán Peninsula in Mexico, ornithologist Dr. Miguel A. Gómez Garza came across parrots with a completely different colour pattern from other known species, and a study published in June, in journal *PeerJ,* names these birds as a new species based on its distinctive shape, colour pattern, call and behaviour. The paper compares and contrasts the distinguishing features of this species with many other parrots.

The new parrot (*Amazona gomezgarzai*), referred to as the blue-winged Amazon because of its primarily blue covert feathers, is characterized by its unique green crown that contrast to blue in other Amazon parrots. This new parrot occupies a similar area in the Yucatán Peninsula as the Yucatán Amazon (*A. xantholora*) and the White-fronted Amazon (*A. albifrons nana*) but it does not hybridize with them.

A very distinctive feature of the new taxon is its call, which is loud, sharp, short, repetitive and monotonous; one particular vocalization is more reminiscent of an Accipiter than of any known parrot. The duration of syllables is much longer than in other Amazon parrot species. In flight, the call is a loud, short, sharp and repetitive yak-yak-yak. While perched, the call is mellow and prolonged.

This species lives in small flocks of less than 12 individuals. Pairs and their offspring have a tendency to remain together and are discernible in groups. Like all members of the genus Amazona, this parrot is a herbivore. Its diet consists of seeds, fruits, flowers and leaves obtained in the tree canopy.

Source: https://www.sciencedaily.com/releases/2017/06/170627073607.htm

New species of Antbird discovered in Peru

A new species of antbird was discovered in Peru last July, by Josh Beck, and the bird has been seen on a few subsequent occasions, including by Gunnar Engblom of Kolibri Expeditions.

Gunnar says: "Peru never ceases to come up with surprises. A new Antbird species was discovered in July last year by Josh Beck near the spot for Scarlet-banded Barbet, less than a kilometer from the village of Flor de Café, aka Plataforma. The Antbird is completely terrestrial and very similar to the Ferruginous-backed Antbird (*Myrmeciza ferruginea*) of Venezuela and the Guyanas.

"In 1995 on a remote peak of the Cordillera Azul a new, remarkable and beautiful Barbet was found during a Lousiana State University (LSU) expedition. The Scarlet-banded Barbet was described to science in 2000. While there certainly was a lot of interest by birders to try to mount an expedition to see the species, very few actually could endure the hardship involved. One needed to set off a full 9-10 days up the Cushabatay river to see the bird and it included a very difficult and steep two day hike up the hill known as Peak 1538 (standing 1538 meter tall) to get to the only area where the bird was known from. Other expeditions into Cordillera Azul, had not produced any sightings.

Source: http://www.rarebirdalert.co.uk/v2/Content/New-species-of-Antbird-discovered-in-Peru.aspx?s_id=251355362

Record migration and breeding season

Record-breaking 120,000 Ruffs counted in Belarus

Thousands of ruffs (*Calidris pugnax*) descended upon Turau Meadow, Belarus in April, breaking records in the country.

During the season, Turau Meadow becomes

a collection point for as many as 150,000 Eurasian wigeon (*Mareca penelope*) and 20,000 Black-tailed Godwit (*Limosa limosa*) which can gather in these plains, sometimes in a single day.

Source: http://www.rarebirdalert.co.uk/v2/Content/Birdlife-Record-breaking-120%2C000-Ruffs-counted-in-Belarus.aspx?s_id=251355362

Isle of Tiree black-tailed godwits

Black-tailed godwits (*Limosa limosa*) have arrived on the Isle of Tiree amassing a record breaking 2,270 this spring, the highest number thought to have ever been counted in Scotland at one time. These large wading birds often stop off in the Inner Hebrides, Argyll, in April and May to refuel during their long migration to Iceland, where they breed.

Tiree typically only sees a few hundred godwits arriving to feed around the loch edges and wet grasslands. The previous record was 1,320 birds back in 2013. The new record, set in April 2017, almost doubles that, representing some 5 per cent of the entire Icelandic breeding population. One of the flocks was spotted on an RSPB Scotland reserve, but the largest – totalling 1,750 birds – was recorded in a tiny field at

Kilmoluaig.

At least 20 individuals were seen with coloured rings on their legs, which revealed that they had spent the winter in a range of diverse regions from France and Portugal to England and Spain.

SOURCE: http://www.rspb.org.uk/community/placestovisit/islay/b/weblog/archive/2017/04/27/record-scottish-count-of-black-tailed-godwits-on-tiree.aspx

Record breeding season for the endangered Bermuda Petrel

The rare Bermuda petrel (*Pterodroma cahow*) is nesting in record numbers this year on Nonsuch Island, where a translocation project has established a viable population in predator-free surroundings. A census earlier this month found a total of 117 breeding pairs (counted as those producing an egg, whether it hatched or not), up by two pairs compared to last year. By mid-April at least 62 chicks were present, so that, even with one or two losses before fledging in late May and early June, it seems likely the previous record of 59 fledged chicks in 2014 will be broken.

Endemic to its eponymous island, Bermuda Petrel — or Cahow — is thought to have been an abundant breeder throughout Bermuda in historical times. Settlement of the archipelago at the beginning of the 17th century saw a drastic population decline attributed to habitat loss, exploitation and predation, and the species was believed extinct for three centuries until it was rediscovered in the first half of the 20th century. In 1951, a tiny relict population of 18 pairs was found breeding on rocky islets in Castle Harbour, and intensive management over subsequent decades resulted in the overall population reaching 71 pairs with 35 young fledged in 2005, when the total number of birds was estimated at 250.

After the breeding season, birds move out into the Atlantic, following the warm waters of the Gulf Stream as far north as the Bay of Fundy in south-easternmost Canada.

SOURCE: http://www.birdguides.com/webzine/article.asp?a=6333

Other News

New Zealand's mainland yellow-eyed penguins face extinction unless urgent action taken

Yellow-eyed penguins (*Megadyptes*

antipodes) - classified as endangered by the International Union for Conservation of Nature (IUCN) - could disappear from New Zealand's Otago Peninsula by 2060, latest research warns, and researchers have called for coordinated conservation action. In a newly published study in the international journal *PeerJ*, scientists have modelled factors driving mainland Yellow-eyed penguin population decline and are calling for action to reduce regional threats. Researchers say that the breeding success of the penguins will continue to decline to extinction by 2060 largely due to rising ocean temperatures. But these predictions also point to where our conservation efforts could be most effective in building penguins' resilience against climate change.

The authors conclude that "now we all know that Yellow-eyed penguins are quietly slipping away we need to make a choice. Without immediate, bold and effective conservation measures we will lose these penguins from our coasts within our lifetime."

SOURCE: https://www.sciencedaily.com/releases/2017/05/170516090847.htm

Rosier future for world's rarest duck

The world's rarest duck, the Madagascar

pochard (*Aythya innotata*), now has a more hopeful future after the Government of Madagascar pledged to protect the wetland earmarked as its new home, and a recent WWT audit of Madagascar's wetlands identified Lake Sofia as one of the only wetlands in Madagascar in a natural enough state to potentially release captive-bred pochards in future.

Lake Sofia's designation as a Ramsar Wetland, also supports the 10,000 people who live in the catchment, the majority of whom live below the international poverty line and are dependent upon the lake for their resources.

SOURCE: http://www.rarebirdalert.co.uk/v2/Content/WWT-Rosier-future-for-worlds-rarest-duck.aspx?s_id=86469103

Gentoo penguin

According to a new study published in *The Auk* (19 April 2017) for the first time, researchers have used time-lapse cameras to study Gentoo penguins (*Pygoscelis papua*) during the cold, dark months of the Antarctic winter. These birds are of particular interest to scientists because their numbers are increasing at the southern end of their range in the Western Antarctic Peninsula, a region where other penguin species are declining. Little is known about the Gentoos' behaviour during the non-breeding season, so a team of researchers used time-lapse cameras to examine patterns in the penguins' presence at breeding sites across their range during the winter season.

Cameras were deployed at seven sites including Argentina, Antarctica and several islands. Each camera took eight to 14 photos per day, and volunteer citizen scientists were recruited to count the penguins in each image via Penguin Watch: https://www.penguinwatch.org/

Researchers found both temporal and spatial factors driving winter attendance, for example, more Gentoo penguins were present at breeding sites when there was open water or free-floating pack ice than when the shoreline was iced in, and more penguins were at breeding sites earlier in non-breeding season than later.

SOURCES:
https://phys.org/news/2017-04-time-lapse-cameras-unique-peek-penguins.htm

https://www.sciencedaily.com/releases/2017/04/170419091632.htm

Wandering albatrosses
A new study has suggested that the body condition of male wandering albatrosses (*Diomedea exulans*) will play a big part in the species' continued survival.

One of the predicted consequences of climate change is a shift in body mass distributions within animal populations. Yet body mass, an important component of the physiological state of an organism, can affect key life-history traits such as survival, chick mass and breeding success and population dynamics.

It is widely expected that variation in body mass distribution will have consequences for the conservation of particular species.

The study concluded that there was a clear difference between the males and females. Increased male body mass enhanced performance in adult survival, breeding success, chick mass and juvenile survival, whereas this was not the case for females.

Adult males also had heavier sons but not heavier daughters. This suggests that a higher investment by fathers in their sons but not in their daughters can increase their overall fitness.

SOURCE: http://www.birdguides.com/webzine/article.asp?a=6359

Research Radar reveals steep declines in Kauai's seabird populations
A new study from *The Condor: Ornithological Applications* has concluded that populations of two endangered seabirds, Hawaiian Petrel (*Pterodroma sandwichensis*) and Newell's Shearwater (*Puffinus newelli*), have exhibited worrying declines on the Hawaiian island of Kauai in recent years.

These species are nocturnal and nest in areas which are hard to access, so monitoring them is challenging. However, using radar to monitor their movements has given researchers a solution. To assess the population trends and distribution of the birds in recent decades, André Raine of the Kauai Endangered Seabird Recovery Project and his colleagues examined past and contemporary radar surveys as well as data on the numbers of shearwater fledglings rescued after being attracted to artificial lights.

Their results shows continuing population declines in both species over the last twenty years — a 78 per cent reduction in radar detections for Hawaiian Petrels and a 94 per cent reduction for Newell's Shearwaters, with the shearwater decline mirrored in decreasing numbers of recovered fledglings over time.

SOURCE: http://www.birdguides.com/webzine/article.asp?a=6412

Database takes record-breaking 10 million seabird tracking points

Seabirds undertake some of the most incredible migratory journeys in the world, and protecting such highly migratory bird species poses a massive challenge. Different scientists, institutions or non-governmental organisations (NGOs) gather local data and try to safeguard their own patch of ocean with limited funds. The Seabird Tracking Database, one of the largest conservation collaborations in the world, was established by BirdLife in 2003 to correlate all the information being gathered around the world on seabirds. It began when data on 16 species of albatross and petrel was put together for the first time to identify the most important places for them and to ensure their protection.

From albatrosses to penguins, petrels to gulls, the tracking database now gathers seabird data from more than 120 research institutes (including BirdLife and its partners) and more than 170 scientists. Data on Critically Endangered species such as Tristan Albatross and another 36 globally threatened species are regularly registered. In total, the database holds information for 113 species, now in more than 10 million locations.

SOURCE: http://www.birdguides.com/webzine/article.asp?a=6416

'Surrogate' hawk mothers orphaned baby birds at Pacific Wildlife Care in Morro Bay

Fiona is a non-releasable red-shouldered hawk (*Buteo lineatus*), who – for the past ten years - has served as a "surrogate" mother to abandoned baby birds taken in at the Pacific Wildlife Care.

She is non-releasable due to being born with a deformity and has been with PWC since 2007. Every year she lays two to three eggs, but they're infertile since there is no male hawk with her. Fiona doesn't realize they're infertile and expects them to hatch. This is why the team at Pacific Wildlife Care decided to swap out the eggs with baby hawks without parents, according to Kelly Vandenheuvel, a rehabilitator and educator for Pacific Wildlife Care.

"What I do each year, when the babies are given to me, is slip them under Fiona just before it gets light in the morning," she said. "I also take her eggs, and place baked egg shells under her so she thinks the babies hatched on their own once it is light, and she can see them."

Fiona is affectionate to each baby bird that comes into her life, as if it were her own, which is precisely the point. Vandenheuvel says she becomes protective, a shelter to the babies, and is noticeably happy to be with her temporary offspring.

"(She's) super protective but as soon as they get their feathers, she will teach them to hunt," Vandenheuvel said. "But when the day comes, she's finished, she's over it. Last year when no orphaned babies came, she sat on her two eggs diligently for three months before I finally had to remove them and take her off her nest," Vandenheuvel said. "It was pretty sad."

In total, she has cared for 14 orphaned birds in a long 10 years.

SOURCE: http://www.ksby.com/story/35417392/hawk-surrogate-mom-orphaned-birds-pacific-wildlife-care-in-morro-bay

There and back again: satellite tagged Turtle Dove returns home

RSPB scientists have succeeded in mapping the complete migratory route of a British turtle dove (*Streptopelia turtur*) - named Lawford after the Essex village where he was fitted with his tracker last summer - for only the second time, after a bird fitted with a tracking device in East Anglia last summer arrived back in the UK on5th May.

After a layover in Croydon and short detour via Suffolk, Lawford made his way to within two miles of where he was first found in 2016, adding to evidence that is helping scientists understand the importance of Turtle Doves' faithfulness to their established breeding territories.

The Turtle Dove is Europe's only migratory dove. Every autumn they cross the Mediterranean Sea and Sahara desert to reach their wintering grounds in Africa, returning to their European breeding grounds in the spring.

Since leaving Essex in September last year, Lawford has travelled over 6,000 miles, stopping in (or flying over) six other countries: France, Spain, Morocco, Western Sahara, Mauritania and Mali.

You can follow the journeys of Lawford and the other satellite tagged Turtle Doves on the RSPB website www.rspb.org.uk/turtledovetracking

To learn more about Operation Turtle Dove and what you can do to help Turtle Doves in the UK, please visit operationturtledove.org

SOURCE: https://www.rspb.org.uk/our-work/rspb-news/news/441350-there-and-back-again-satellite-tagged-turtle-dove-returns-home

Vultures smear their faces in red mud which they use as makeup

Putting on make-up is a rare phenomenon in birds - known as cosmetic colouration - but Egyptian vultures (*Neophron percnopterus*) have been filmed doing this for the first time. Normally sporting a yellow wrinkled face surrounded by a halo of white hair, many of the vultures on Fuerteventura island in the Canaries off the coast of Africa, sport reddish heads and necks, with the colour varying from pale brown to deep crimson. These vultures dip their heads in red soil and swipe from side to side, carefully dyeing their head, neck and chest red. It is a well-studied population, so almost every vulture on the island is marked with plastic rings, allowing researchers to study individual

differences in this curious behaviour. "It's the first documentation of this behaviour in wild birds that are individually marked," says Thijs Van Overveld of Doñana Biological Station in Spain. "The most interesting part of our observation is that there is great variation among individuals in the extent to which they paint feathers, ranging from almost completely white to almost completely red," he continued. The vultures did not follow a particular pattern while mud painting and the baths were not restricted to a particular age, or sex.

Although the related bearded vulture (*Gypaetus barbatus*) is known to display a similar behaviour as a signal of dominance, the researchers don't believe that Egyptian vultures paint themselves for this reason. Unlike the Egyptian vultures in the Canaries, the bearded vulture goes for its mud baths in secret, and the mud daubing itself is a lot more elaborate.

One possible explanation for this behaviour is that the mud keeps bacteria and viruses away. But, if bathing had such a big advantage, many more birds should be taking long mud baths. The authors believe instead that the painting serves a visual rather than health-related purpose, "given the great effect on the general appearance of these otherwise white birds."

Robert Montgomerie from Queen's University in Ontario, Canada, thinks so too. "The amount and incidence of reddish head plumage on Egyptian vultures is quite rare, suggesting that those who have it are indicating something special," he says. But it remains to be seen what that function might be.

SOURCE: https://www.newscientist.com/article/2130980-vultures-smear-their-faces-in-red-mud-which-they-use-as-makeup/

For those of you not aware, as well as this column in *Animals & Men,* Corinna writes a daily Fortean bird blog which can be found as part of the CFZ Blog Network, but also as a stand alone site at:

http://cfzwatcheroftheskies.blogspot.com/

A PECULIARLY 21ST CENTURY BUTTERFLY CONUNDRUM

Regular readers of my inky fingered scribblings, here and elsewhere, will probably be aware that I am fascinated with the comings and goings of British butterflies. Indeed, in these very pages, I have, over the years, presented a number of round ups of the latest information on the subject. And this – on face value - could be seen as more of the same. However, although it is, indeed, a round up of a number of sightings of out of place butterflies in Britain, it has a new and – somewhat disturbing – angle to it.

My two main sources for interesting sightings of British butterflies are Adrian Root's excellent Bugalert website and the UK Butterflies website, which has basically become the industry standard reference source for information on British butterflies.

During the second half of June, there were sightings of 3 species not normally known from the United Kingdom. The first of these was the Camberwell Beauty (*Nymphalis antiopa*). Despite its very English sounding popular name, this has never been a resident of the UK, at least since records began about 300yrs ago. It is, as the name implies, a Nymphalid, and

the common name comes from a description by Moses Harris in *The Aurelian* published in 1766, based on two individuals that had been caught near Camberwell in 1748.

L. Hugh Newman noted in his book *Living with Butterflies* (1967) that this is an apparently non-migratory species from Central Europe and Scandinavia (although it is also found widely in North America and other parts of Eurasia), and that Camberwell Beauty catches in England were suspiciously concentrated around London, Hull and Harwich, all these being ports in the timber trade with Scandinavia, and theorized that they had hibernated in stacks of timber which was then shipped to England, and had not travelled naturally.

This makes this species a fascinating one, because although human beings appear to be the vectors of its introduction into non-native areas, the human beings' actions are completely involuntary. Despite this, its appearance in Britain - and therefore on the British list - can be seen as completely natural.

Bugalert describes this species as a "scarce immigrant, though almost annual", and

JONATHAN DOWNES

goes on to say that it is most often reported during April, August and September. So, the following three records, which are far enough apart from each other to preclude them being the same individual, would seem to be highly atypical:

- 2017.06.24 Bedfordshire Maulden Wood 1 Beds & Northants BC
- 2017.06.23 Norfolk Bungay (nr) 1 Brazil, Andy
- 2017.06.17 Sussex Cissbury Ring 1 Sussex BC

The second species is the Swallowtail (*Papilio machaon*).

The British race of this widely distributed species is restricted to the Norfolk Broads, where it emerges between May and mid-July, with an occasional second emergence in August. The following sightings were recorded by Bugalert between May and June:

- 2017.06.19 Norfolk Potter Heigham Marshes 15+ Mills, Chris
- 2017.06.14 Norfolk Strumpshaw Fen 2 Jenkins, Gareth
- 2017.06.14 Norfolk Upton Fen 2 Jenkins, Gareth

- 2017.06.17 Berkshire Yattenden 1 Upper Thames BC
- 2017.06.17 Essex Colchester Community Garden 1 Cambs & Essex BC
- 2017.06.17 Kent Northdown Park, Margate 1 Rare Bird Alert
- 2017.06.15 Norfolk Upton Fen 1 BDS
- 2017.06.15 Norfolk Hickling Broad 12 Riley, Adrian
- 2017.06.15 Norfolk Potter Heigham Marshes 20+ Riley, Adrian

- 2017.05.25 Norfolk Strumpshaw Fen 5 UK Butterflies
- 2017.05.21 Norfolk Strumpshaw Fen several Norfolk BC
- 2017.05.12 Kent Northdown Park, Margate Ssp gorganus 1 Rare Bird Alert
- 2017.05.10 Norfolk Potter Heigham 1 Norfolk BC
- 2017.05.02 Ayrshire Rozelle, Ayr 1Southwest Scotland BC

As can be seen, the vast majority of these sightings are from East Anglia, and, we feel can be broadly dismissed as being of

the native sub-species in the place and time that it is meant to occur. However, specimens from Berkshire, Kent, Ayrshire, and – less conclusively – Essex, are far more problematic. The first of the two Kent specimens has been identified as belonging to the mainland European sub-species. This was once a British resident, and evidence from the past few years suggests that it may become so again. But, whereas sightings of these butterflies along the south coast of England are quite likely to be naturally occurring vagrants, ones inland are again more problematic.

The third species of interest in this roundup is undoubtedly one of the most iconic butterflies in the world. The Monarch (*Danaus plexippus*) is essentially a North American species, which carries out enormous migrations each year. It spends the warmer months in the Northern United States and Southern Canada, but every autumn enormous numbers of this butterfly migrate southwards to overwintering grounds in California and Mexico.

I lived in Canada for a few months during the summer and autumn of 1979, and the sight of massive trees with their branches weighted down almost to breaking point, by the sheer biomass of specimens of these butterflies preparing to fly south, is something that I will not easily forget.

Over the years, they have spread to many parts of the world, where they were not originally found and non-migratory populations are now found in Spain, Portugal, Indonesia, Australia, New Zealand, North Africa, Hawaii, the

Caribbean, the Azures, the Canary Islands and Madeira, as well as an artificially propagated population on the International Space Station. In the UK, they turn up most years, usually in the far south west, and usually during the autumn.

However, on the 18th of June, one was seen and photographed in a cemetery in Rochdale.

What are these three sets of records got in common? Well, all three species are on the British List, but there are indications (as we have shown) that none of the sightings listed (except for the East Anglian Swallowtails) are conclusively natural in origin.

As we understand it, it is an offence under the Wildlife and Countryside Act to knowingly release non-native species into the British countryside. But, we have discovered, this is exactly what one of the most important hidden industries in the United Kingdom does on a regular basis.

I last read Jessica Mitford's *The American Way of Death* (1963) some years ago, but it had a major effect on me.

Feeling that death had become much too sentimentalized, highly commercialized, and, above all, excessively expensive,

Mitford published her research, which, she argues, documents the ways in which funeral directors take advantage of the shock and grief of friends and relatives of loved ones to convince them to pay far more than necessary for the funeral and other services, such as availability of so-called "grief counsellors," a title she claims is unmerited.

On a personal level, I find the crass commercialisation of the way that we mark important events in our lives, and deaths, to be somewhere between irritating and immoral, and when I have been in the position of having to arrange such things I have always done them as cheaply and unostentatiously as I can. However, I am in the minority. I have recently discovered that various people that arrange wedding ceremonies and – indeed – those who arrange funeral ceremonies, particularly those of children, have started a practice of releasing butterflies at such events, and charging an enormous amount of money so to do.

When I told my friend and colleague Lars Thomas about this, he said: 'Egad! Why don't they just release giant Japanese hornets and vampire bats?', and one has to agree with him.

Just a cursory perusal of the websites of

some of the companies offering these services will confirm that the three species I have listed above are amongst the most commonly released as part of obsequies or nuptials, organised by people who are prepared to pay £150 for ten insects.

Until now, it has always been thought that because various plants of the genus *Asclepias* (milkweeds) do not grow naturally in the United Kingdom, and can only be grown in greenhouses here, that monarch butterflies will never become naturalised here.

However, in recent years, various species of milkweed have become widely available in garden centres and even supermarkets, and will now grow – at least in the summer – in British gardens. In addition to that, I have been informed that initial research by one of Britain's major zoos suggests that, under certain circumstances, monarch caterpillars will eat a wide range of other food plants. Therefore, the above cited reasons why these butterflies cannot become nationalised residents of the United Kingdom would seem to be becoming increasingly spurious.]

This is only a position document to explain where we stand at the moment with an ongoing research project. We have our Lancashire representative canvassing wedding planners and funeral directors to find out how prevalent these practices are in her area.

I also have other CFZ operatives trying to surreptitiously find out what the official word is from Defra on the subject. I do not want to be in the position of stopping new species of British resident butterflies from taking a hold here. God knows that our biodiversity is in enough trouble as it is. But, we feel that it is the job of the Centre for Fortean Zoology to look further into such occurrences, particularly when they are as bizarre as this one.

Watch this space.

Available Butterflies
(pictured approx. to scale (may differ on small screens))

Monarch
Wingspan approx. 4"

Swallowtail
Wingspan approx. 3½"

Red Admiral
Wingspan approx. 2½"

Painted Lady
Wingspan approx. 2"

Mourning Cloak (Camberwell Beauty)
Wingspan approx. 3½"

THE LONG AND WINDING TOAD

I recently came across a story via the Early English Books Online web site which was titled `A true narrative of a strange prodigious toad for size and shape full fourteen inches long, and ten over: the toad on the hinder-parts: with a tayl three times as long as his body, with a forked sting at the end.: Seen lately at a hunting about the Devil`s -Arse-oth`-Peak in Derbyshire: The sculpture of which, in copper , will shortly be published, and examin`d as to both more narrowly by some who are now in the city and whose usual curiosity in things so rare, led them to this exactness of observation.

A true narrative of a strange prodigious toad for size and shape full fourteen inches long, and ten over : the toad on the hinder-parts : with a tayl three times as

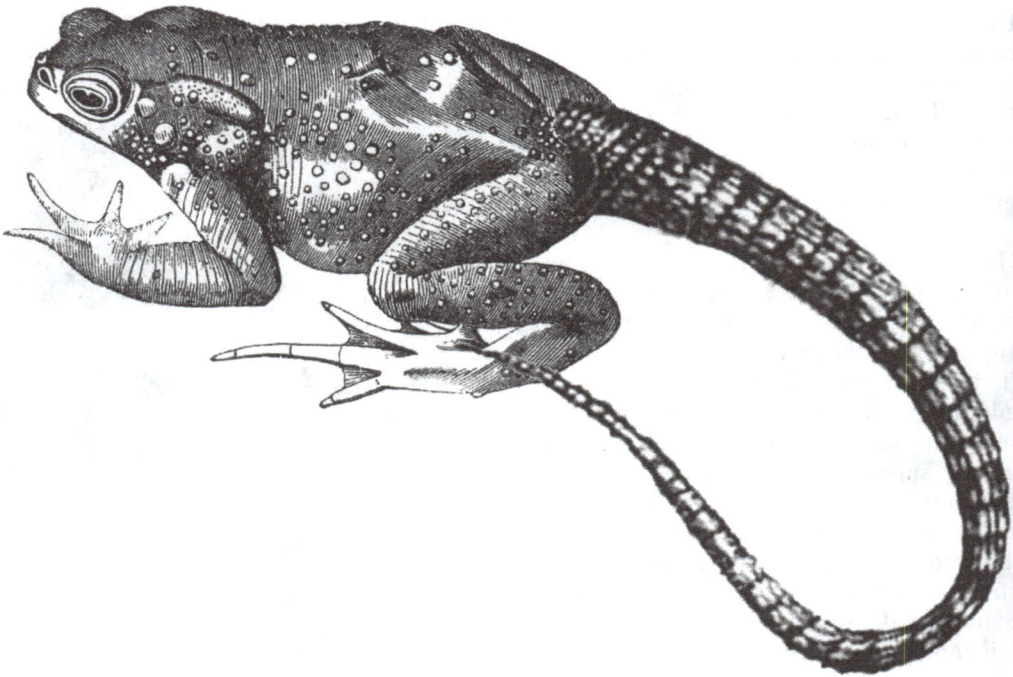

RICHARD MUIRHEAD

long as his body, with a forked sting at the end. : Seen lately at a hunting about the Devils-Arse-oth'-Peak in Darbyshire. : The sculpture of which, in copper, will shortly be published, and examin'd as to both more narrowly by some who are now in the city and whose usual curiosity in things so rare, led them to this exactness of observation.L'Estrange, Roger, Sir, 1616-1704.

There is nothing so common as Natures variation through luxury or poverty of matter, and some concurrent accidents, which taking the opportunity of time and place, assist in producing such an Animal Mineral, or Vegetine, as because of it's rarity will seem strange to the world.

For Barnicles some years ago, nay now, are not believ|ed by the ninth part of ten of the Nation, though several Authours have writ, and many thousands seen the manner nature has contrived for them; they therefore that can have little belief for any thing, of which they have not had an occular demonstration are not expected to give cre|dit to this ensuing Relation.

HAlf a Dozen of the meaner sort of Darby|shire designed a Hunting, Coursing or Potching rather, having for that purpose got Beagles, Grey-hounds, and a Peece or two, thus furnished, they beat about the Peak, where they started a Hare as some held, others a Rabbet, which they eagerly followed, until that frightned Crea|ture ran into a hollow in the earth; at the entrance of which stood the Dogs, not daring to go in; but when the company came together, and had consul|ted a while, they resolved to enter and drive out the prey, the cavity being large enough, the most couragious went in a good way, but one whose heart or weariness had caused to stay behind, sat him down on a great stone loose and moveable; and whether by fitting down he stirred the Stone, he presently conceived his fellows were gone too far, he called them back, telling them, that what they looked for was under the stone he sat on: up|on which they returned, carrying a Beagle or two in, which durst not voluntarily approach, then ve|ry circumspectly they turned the stone, but found their expectations jilted, for which they called the Informer a silly Coxcomb, who had hindred their further search; but he confidently aver'd he per|ceived something move, and laying his hands an|gerly upon the stone, laid his fingers upon a round hole in't, into which he try'd to thrust his hand, but failing, cry'd, he would lay his Soul to a Pot of Dar|by-Ale, that the prey got in at that hole, upon which they all try'd, and one whose hand was slenderer, thrust it in, arm and all, and finding his utmost fin|ger to touch a hairy live body, swore bloodily, and curst himself to the Devil, it was a Rabbit, for he felt it; upon which the Inventor reply'd, I knew I could not be deceived, then they all agreed to roul the stone out, being near to the mouth, but they were never the better to further their design.

The most ingenious of them cut a Bryer, slitting the end of it a cross, and afterwards cut it to point towards the middle, and gently thrust it in, until he found it stopt upon something that was soft, then he began tenderly to turn it, and drew out a hoary furr, which they concluded to be the belly of the Rabbet; after many essays, finding themselves dis|appointed, they resolved rather then loose their ex|pectations, to get some Iron Mawls and Barrs, and so break the stones to pieces; which they borowed, and presently fell to Work, and after considerable pains they split the stone, in the midst of which to their amazement and great disappointment, they found a Toad of prodigious shape and size.

After their astonishment was something over, their an|ger took place, which they intended to vent against the in|nocent poor creature; when by meer accident a Stranger came by, and seeing in what confusion they stood, drew near and was equally surprized, but finding that they had prepared to brain the poor Creature with part of his own Castle, entreated them to forbear, and to recompence their pains in some measure, he would gratifie their Civility with a Dozen of Ale, which they gladly accepted of, and immediately went to Drink, whilst he had the liberty of viewing this monstrous creature, which he found to have all the parts of a Toad and more; therefore having com|petent skill in Limning, and having the advantage of Pen|cil and Paper about him, did with all exactness Draw this strange Animal, which from his own hands came to mine, whose Description you shall now have, and when the En|graver with whom the figure is, has finished it, that shall be also presented to you.

This Creature was in Length exactly Fourteen Inches: as it Sat it was Ten Inches and a Half over; upon the hinder parts of his Back, and the fore-part eight and a Barley Corn, how high it might stand he knew not, for it sat all the while crooling down; its nose was as black as jet, proportionable to the rest of its body, and of each side adorn'd with Bri|stles; under his Eyes was a perfect Seaming made Cistern-fashion, of a Green colour, whilst his eyes most delicately beautiful, seem'd to drop that way; over which were placed bristles of a muddy green; under his Gills were Flapps like two scollop shells, of the Colour of a Gurnets fins, admirably blew and shining, ꟻ always expanded; under them were two Pouches or Bladders, of a deep Orange colour, larger then the Fins, being a big as ordinary Pear|plumbs, and something of that figure▪ the head had a Sea-Green scaly Armour on, which ran along itsPage 6back within four Inches of its tail; the hinder-part was covered with perfect Hair, but dusky, hoary co|loured, of the same manner was that as went along the sides and feet. But what is yet most strange, is his Tail, which was thrice as long as its body, which it wore on its back, somthing like the True-Loves|knot, and tho' it seem'd so intricate, yet when he of|fred to turn this creature on its back to view its bel|ly, it flung its Tail as sudden as the jirk of a Coach|mans Whip, ⅂ had not he stood sideways, he might have had too much cause to have repented his cu|riosity; but as it hapned it's tail fell into a slit of that stone which was its former Mansion, and though the poor Creature strove hard, yet it could not re|lease it thus engag'd; upon which there appeared as close as might be, onely reckonable, a hundred ninety seaven scaly rings, jetty, black, and shining, which he made a shift by this means to reckon, and may be guess'd to be the indications of its Annual extension; at the end of the Tail was a forked sting, within an Inch of which was a small Pouch, of the colour of Gum Bugiae.

The Gentleman finding his Over-grown Toad in this posture, would have gagg'd the Toad, but no sooner had he touch'd its Nose, but it recoil'd back|wards, even off the Stone, which it seem'd so to Hatch, and Cherish before. But there being an un|fortunatePage 7hollow or miss of ground, where this rare Animal falling, ⅂ not coming quick enough to the ground, the said Satchel or Pouch neer the sting broke, out of which there issued the most Diapha|nous Yellow Liquor as ever was seen, deeply stain|ing the stone, upon

which the Toad that then hung above-ground fell down, his Tail freely coming to him; which he had not enjoy'd above three Mi|nutes e're he departed this scurvy world: but before it received its change, it gave three such shrieks that the stranger was not onely forc'd to stop his ears: But this Lamentable Dirge, forc'd the Neigbours out of their Houses: He perceiving the Creature quite dead, took a fresh resolution, and turn'd it up|on its back, and found its belly to be as white as the driven snow; cover'd with a most delightful down: about that time when he was busie viewing this pleasing Object, his Nose was surpriz'd with a most ungrateful Stench, which he found to proceed from that Limpid Yellow substance that the Toad had spilt, in which (it seems) its Seal of Life was plac'd. Just at its Departure its Head mov'd, and it vomited forth a Triangular Stone, being an Inch from each Angle, and in Depth according, which he drew to him with a Staff, keeping his Fingers upon his Nose: and when he had got it at a Convenient Distance, he Doubled somePage 8Paper and took it up, folding it in several folds of the same Paper, and after lapt it carefully in his Handkerchief, and presented it me, with the Figure: And tho' before, I was in the num|ber of the Incredulous, and so far from giving credit to the many Excellencies of this Stone, that I question'd its Existence, and guess'd it to be onely the off-spring of some pregnant, yet imposing fancy: Yet now upon my short ex|perience, I dare assure the World, that the vertues ascribed by Authors to it, come far short of its performance: And questionless had some curious Virtuosus of our times been present at this Prodigious Accident, they might have pre|pared Medicines of equal Memory with their Fame.

FINIS.

The "toad" or whatever it was in Derbyshire, the eye witness account of which was published by or on behalf of Sir Roger L`Estrange in 1677, is very similar to one that was seen in Mendham, Suffolk by Thomas Flatman in 1662 as described in a letter to his brother dated September 25th 1662. But first an introduction to the Devil`s Arse toad, lizard or salamander. I raise the possibility that is was a salamander because of the mention by L`Estrange (note the Fortean sirname) of it`s hair on some parts of its body: "...his eyes most delicately beautiful...over which were placed bristles of a muddy green;..the hinder-part was covered with perfect Hair, but dusky,hoary coloured..." And so the description continues, as you will see. But yet the cryptid had armour or scales, whereas salamanders and toads are smooth. The whip-like tail is clearly a non-ophidian element. It reminds me of a mature pterosaur or pterodactyl and whilst I am not at all suggesting that we have a pterodactyl (a juvenile?) here in Suffolk or Derbyshire, I am suggesting no contemporary known amphibian or reptile fits this 17th Century mystery beast.But I have never up to now come across as detailed a description of a cryptid whether contemporary or historical as this one.

The image below dates from a time when it was believed by some that salamanders were furry, as in the Derbyshire cryptid,but as Forteans, who are we to say they were wrong? The image here comes from a web site titled `Fantastically Wrong:The Legend of the Homicidal Fire Proof Salamander`

(https://www.wired.com/2014/08/ fantastically-wrong-homicidal-salamander/?mbid=social_fb_onsiteshare)

However I have not been able to find the copper engraving L`Estrange refers to.

I do vaguely recall seeing in a modern reproduction of a medieval bestiary a lizard-like animal which reminded me of the Devil`s Arse (so-called because of the sound of the air moving through the chasm`s opening) "toad". But an important point to remember is, that although L`Estrange was a prolific, almost manic Royalist pamphleteer during the time of the Civil War, whose ambitions led to at least one prison sentence (on death row no less) it could be that the whole story of the toad could be an elaborate satire by him or against him, the exact meaning of which is now lost in the mists of time. As someone calling herself/himself Enola Gaia pointed out on the *Fortean Times* Forum on July 2nd 2017:

1. Was the ...Toad pamphlet not a news item, but rather some sort of veiled /

allusive political tract whose metaphorical references are now lost on us?

2. Was the ...Toad pamphlet one of the pseudo-L'Estrange publications (again, metaphorical / political) against which he had to defend himself?

The Suffolk "toad" (or whatever it was) was described by its witness as being a yard and a half in length, head like a toad, very large yellowish ring around its neck, (like a grass snake) two wings like a bat`s, four legs like a duck`s - short. Dr Karl Shuker in his `The Menagerie of Marvels` (2004) compared it to a *Draco volans* flying lizard.

The image below is of Peak Cavern aka Devil`s Arse, Mike Peel, Wikipedia Commons. This cavern also housed the last ever community of British troglodytes.

Ronald Murphy is an American academic and author who is writing a series of books for the CFZ press. The first of these, *On Vampires*, was published earlier in the year and we are very pleased to bring you this exclusive extract

The Vampire God of Ancient Egypt

Ronald Murphy Jr. graduated from the University of Pittsburgh with a degree in Literature and Religious Studies and attended graduate school at Pitt and at Indiana University of Pennsylvania where he studied history. He is a professional actor, having appeared in movies, on television, and on stage. He is also a researcher and historical reenactor for the Underground Railroad in Blairsville, PA. Ronald is considered an international expert on faerie lore, and has researched the unexplained from Maine to Florida as well as in the United Kingdom. He is a noted lecturer, and appears at various conferences throughout the year. Ronald is interested in infrasound and pheromones in relation to Fortean research. He also studies cryptozoology as it relates to the Collective Unconscious and focuses on archetypes found throughout world cultures

Ancient Egypt is a vast period in our study of the development of the vampire, a culture that spans from as far back as 3100 BC to the time it was conquered by the Greeks in 322 BC. This was a crucial epoch in human progress, and the influences of this culture reverberates in the human consciousness to this very day. As we continue our survey into the historical notions of the vampire, it is important to understand that ancient Egyptian theology was populated by demons. These demons had the attributes of the natural world, embodied in an anthropomorphic deity. The funerary god Ammit, for instance, known as the "devourer of the dead" or "soul-eater," was a female demon with a body that was, appropriately enough, part lion, hippopotamus and crocodile. These creatures represent the three largest "man-eating" animals known to ancient Egyptians. The Egyptians were imaginative, but they were also very practical. Nature was still seen as a chaotic realm imbued with spirits, both beneficial to humankind and extremely malignant. The Egyptian religion was an extension of the animistic beliefs of the antiquity of the human race, now illustrated as gods that personified elements within the natural world. Like nature, the impact of these gods on humanity could be mercilessly vicious.

Shesmu was a god who was very seldomly depicted, but when he was illustrated artistically, he appeared as a man with a lion's head, holding a butcher's knife. Again, we see the idea of the animal as the head or driving force of the human body. Since the time of the Old Kingdom, at the very dawn of the Egyptian civilization, this demon-god Shesmu was a source of fear. You see, Shesmu was very vindictive and bloodthirsty. He was known as the "Lord of Blood," the "Great Slaughterer of the Gods" and "He Who

Dismembers Bodies." Although in human form, Shesmu was a reminder of the animal instincts that still lurk in our DNA. As we have seen in the figure of Lilith, the natural world informed ancient cultures, and it was no different in the Egyptian world. The aspects of man and animal, civilization and chaotic nature impacted the Egyptian culture on how these demons and gods interacted with the human world. After all, the ancient Egyptians lived in a world animated by forces over which he had little control. By naming these forces and giving them shape and character they attempted to find some understanding and ultimately ways to influence their behavior. The Egyptians believed in a pantheon of gods, which were involved in all aspects of nature and human society. Thus this culture's religious practices were ritualized efforts to endure and possibly placate these natural phenomena. Death was one force in nature that the Egyptians obsessed over. And death was intrinsically linked to the concept of the vampire.

Like many cultures that existed in the ancient world, the mystery of death was a preoccupation stitched into the tenuous fabric that held civilization together. According to the Royal Ontario Museum, childbirth was a most perilous event for the mothers and for their babies. In Ancient Egypt, maternal and infant mortality were high, which many researchers believe to have hovered often over a 30% mortality rate. This incredibly high rate of maternal and infant death was not exceptional, however. For most of human history, somewhere between 20% and 50% of babies have not lived through their first year of life. In Ancient Egypt, a mother would be very lucky if half of her children lived to be adults. Many men would lose their wives in childbirth, and many older children would grow up without their mothers. Of course, there were many causes of this high death rate in infants. Many

diseases that can now be cured, such as smallpox, were quite effectively deadly before the advent of modern medicine. According to the information compiled by the Royal Ontario Museum:

All people until very recent times were greatly troubled by parasites, from the lice and fleas that suck human blood, to more subtle and often more deadly creatures that live inside human beings. A water-borne parasite brings a disease called bilharziasis or schistosomaisis, which still affects many people in the world. Tape worms and other parasites could be introduced to a body if food, especially meat, was not cooked thoroughly. Trichinosis is one such disease, caused by eating undercooked, infected, pork. Malaria still kills a million people a year throughout the world; it was present in Ancient Egypt, sapping people's strength, making them more susceptible to disease and infection.

Again, we see the vampyric figure in the ancient world closely associated, and, indeed, even identified with, pestilence and contagion. It would seem, indeed, if some sort of force, a malicious energy, possibly even an evil entity, was preying upon the weak and the vulnerable. Lilith was accused of feeding off of the life-force of newborn children and sucking the life from a mother's breast in cultures throughout the Fertile Crescent. Egypt, too, meted out blame to nebulous forces that drifted in from the desert to prey upon the weak and vulnerable.

Therefore, apotraics were employed to guard against these evil forces thought to be stalking children and mothers. By employing natural magic, the ancients believed they could combat these spirits that seeped into humanity from the liminal zones of the desolate deserts. It would have seemed as if the Jungian shadow had taken form and sought after blood like a parasite, and in a pre-scientific culture, this is exactly what the ancients believed was happening. One such apotraic used was garlic. This healing herb was so revered by the Egyptians that it was discovered buried within the tomb of King Tutankhamun. The Egyptian medical text known as the *Codex Ebers,* prescribed garlic to heal the body afflicted with parasites and insect infestation. Since parasites literally feed off a host, it is fitting that the vampire would develop this sort of one on one correspondence with this blood-sucking creatures. Garlic was also prescribed to treat skin diseases. Certain skin diseases could mean that you were excommunicated from society and forced to live in the barren wastelands, the realm traditionally of not only the unclean but also the domain of the vampire. Indeed, this herb is still a believed weapon against the vampire even to this very day. But why was garlic used to protect newborns and new mothers? Well, it seems, along with garlic's other interesting healing properties, this odoriferous herb also stimulates the flow of breast milk and may, in fact, instigate a newborn into latching onto the mother's breast after a new mother has ingested garlic. A newborn latching on to feed and a new mother producing enough milk is, of course, essential to the life of the child. Therefore, the child has a better chance of surviving infancy. It is my contention that garlic was seen as a talisman against the vampire's predations of infants and mothers because the well-fed child would not be seen to wither away and ultimately die as if it was being slowly preyed upon by some mysterious force in the night. Therefore, the garlic was believed to be the salvation of the child, the bane against the predatory vampire. Eventually, this protection by garlic lost its association with childbirth and has come to us as a tradition of being effective against the vampire.

Because life was fraught with so many dangers, both real and imagined, in the guise of not only evil spirits but in the form of virulent disease and natural disasters such as drought or the seasonal flooding of the Nile, as well as from animals such as snakes, lions, and even the hippopotamus, death was a constant shadow within this culture. Death loomed over the ancient Egyptian like a tangible specter, a shadowy presence from the moment of birth, ominously following the individual in childbirth and throughout adulthood. It is no wonder that the afterlife was such a vital belief in the Egyptian religious beliefs. The afterlife was something promised after the uncertainty within the living world. Because the Egyptians were so preoccupied with the afterlife, it is only fitting that the idea of the vampire was beginning to take a concrete shape in their mythology. Egyptian views on death and the afterlife were complex, and the treatment of the deceased was of vital importance. As we have seen in the Neolithic burials where caves were used to inter the dead, caves too were used in Egypt. However, to hold the eminent members of Egyptian society, artificial caves made out of mud bricks were erected. These edifices are called mastabas and first appear at the very beginning of Egypt's history, in the period known as the Old Kingdom. The mastabas mimicked caves in that there was an above ground chamber that led to other subterranean chambers. As the funerary technology advanced, pyramids would replace the mastaba tombs, however the pyramids simply expanded on the same principle on a monumental scale. The first pyramid in Egypt was constructed as early as 2540 BC, forming an artificial mountain with a cave-like entrance and enclosed burial chambers. Even the Great Pyramid at Giza also has chambers, reminiscent of the Cheddar Caves in England. As ancient man used caves for burials in the Stone Age, the Egyptians created their own receptacles for the human spirit. After all, the proper preservation of the body was essential for passage into the next world. This was the reason that mummification was such an intensive and painstaking undertaking. It is in this facet of ancient Egyptian belief involving the life-force known as the ka that prompts vampyric consideration. The ka was thought to be an astral companion to each living person. The ka was a symbol of the life powers given to each man from the gods. The ka was said to be a spiritual double that was born with every man and lived on after he died, but only if it had a place to live. The ka lived within the body of the individual and therefore needed that body after death. This ka, this essential vital spark, is the essence that distinguished the living from the dead. When a person died it was believed that the ka left the body. Food and drink was given to the dead so that the ka could feed from the essence contained within the offerings. We now see the beginnings of the vampires tradition of an undead force needed sustenance in order to survive. But the ka informs the vampyric tradition further, for the reason for the extensive and elaborate preparation of the dead body was to ensure that this ka had a home. The belief was that the body, even after death, had to resemble the living body as much as possible so that the ka could recognize it.

This was the reason that mummification was such an intensive and painstaking undertaking. First you had to make sure the body would not rot, so a hook was inserted through the nose to remove the brain. You also had to remove all the internal organs, but the lungs, intestines, stomach, and liver were important in the afterlife so these organs were stored in containers called canopic jars. These jars would be placed in the tomb with the mummy so the mummy could use them at a later time. The heart was placed back inside the body, then the corpse was rinsed with wine and various spices. Then natron, a kind of salt, would cover the body, pulling all the moisture

from it so the body would not decay. This process took 70 days to complete. Then the body was stuffed with linen or sand to give it the shape of a living human. At the end, the body was wrapped from head to toe and placed in a sarcophagus. You see, it was also believed that the ka would leave the dead body at venture into the world of the living. There was always the Egyptian fixation of helping the dead on their journey to get to the next life and to assure that the spirit of the deceased was contained outside of the world of the living. As long as the body was properly prepared and buried with the appropriate rites and the deceased's name was continually remembered, the spirit would journey well and rest in eternal peace. However, if any of these conditions were not met, it was believed the spirit of the deceased would walk the earth and wreak havoc, causing nightmares and illness. As you can see, the human vampire rising from the tomb was developing in the Egyptian world.

The idea of the vampire was also deified in the pantheon of the Egyptians as well. The Egyptian gods were anthropomorphic personifications of the forces of nature, and the vampire was definitely considered a natural element of the world. In Egyptian mythology, Sekhmet was the vampire goddess. Sekhmet appears early on the Egyptian stage, possible at the very onset of the Old Kingdom period at the very dawning of the civilization, over 5000 years ago. Sekhmet is a goddess who is primal, alluding to the conditions of life before the Egyptian culture organized and formed their society along the waters of the Nile. This goddess is the cultural memory of life before the barriers of society that demarcated the civilized from the chaotic natural world. Sekhmet was also viewed as a warrior goddess as well with a taste for blood. This goddess was a creature of animal instinct and bestial territorialism. Appropriately, Sekhmet was depicted as a woman with the head of a lioness, the fiercest hunter known to the Egyptians. Even at the zenith of the Egyptian culture, lions were respected and feared. As in areas of the world today, lions still preyed upon humans. Sekhmet, like the lion, was untamed, living just beyond the reach of human culture. Indeed, it was said that Sekhmet's breath formed the desert, the liminal region on the outskirts of human influence. The desert was the realm of pestilence and disease, contagion and malevolent spirits. Moreover, it can be speculated that the desert regions were also equated with the realm of the dead as this was the area allotted for burials. We see this idea blatantly illustrated in the artistic representation of Sekhmet, in fact. The lion head of the goddess is often depicted as painted with a hue of green. The color green in Egyptian iconography was the color of the dead as it was the natural shade of decomposing flesh. So Sekhmet was the goddess of the liminal regions assigned to the dead because she was one of the dead.

Sekhmet was also known as the bringer of death. Sekhmet's name comes from the Ancient Egyptian word "sekhem" which means "power or might." Sekhmet's name suits her function and means "the one who is powerful." Sekhmet is primal, retaining the animistic drives as the demons that haunted the Egyptian worldview. But she is also a figure like Lilith. She is powerful and foreboding, a female force that cedes to no one. In this capacity, she was also given the pleasant and reassuring titles such as the "Mistress of Dread," "Lady of Slaughter," and "She Who Mauls." So, as we can plainly see, this goddess is not a dainty debutante. No, Sekhmet bursts into the consciousness with as much civility as a lunging jungle beast. But she is also depicted as having human components as well. Sekhmet is often shown in Egyptian art as wearing red clothing, an overt allusion to blood, to cover herself from the waist down. This may be a

representation of menstruation, encapsulated in the image of this goddess. After all, Sekhmet is not totally animal; indeed, she is also a vital woman, bare-breasted and nubile. Like Lilith, Sekhmet is a dichotomy of fear and attraction, dread and throbbing anticipation. Sex and death are further intrinsically linked in this Egyptian goddess. Fittingly, another one of Sekhmet's symbols is the cobra, an appropriate symbol of temptation adapted by other religions in this area as well. And Sekhmet does tempt us, inviting us in. But if we get too close, overcome with lust and desire, the figure of the cobra reminds us of the deadly sting that awaits us. Beware of the fangs that lie in wait, hidden, ready to strike out at the unwitting.

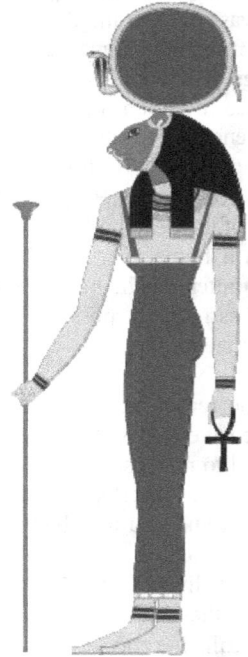

A reproduction of an Ancient Egyptian depiction of Sekhmet.

Sekhmet is a complicated character. She is

dangerous yet alluring. Sekhmet is thus an illustration of the dichotomy inherent in the human condition. Like the other Egyptian gods and goddesses, the figure of Sekhmet is a mirror. In her we see our reflection. She defines who we are because the Egyptian gods were made in our image. They encapsulate the human condition and represent the forces of the natural world interacting with the humanity. Sekhmet is at once a refined goddess and a devourer of the living. We, as a human race, may seem little less than gods, but try as we might to forget our savage past, we all have cannibals in our closet.

As such a primal, bestial character, Sekhmet was undoubtedly feared by the Egyptians. Sekhmet was a deity who presided over the unpleasantness of war, licking her lips as blood spilled on the ground. Because a lioness was perceived by the Egyptians to be the fiercest hunter, it was appropriate that it was believed that this goddess led the pharaoh into war. At the end of battles, festivals were celebrated in hopes to pacify Sekhmet so that the destruction and killing would come to an end. War in any culture is unpleasant, a seemingly living organism that takes the lives of the young in a bloody feeding frenzy. Like other cultures in the ancient world, propitiation in the form of sacrifices were offered in the hopes that Sekhmet's blood-lust would be quenched.

But this goddess Sekhmet was always on the prowl, stalking in the forces of nature, uncontrolled and unpredictable. Thus, in this manifestation of natural forces, Sekhmet also needed to be appeased yearly as well. During an annual festival held at the beginning of the year, which was essentially a feast of unbridled intoxication, the Egyptians danced and played music to soothe the wildness of the goddess and drank great quantities of wine ritually to imitate the extreme drunkenness that stopped the wrath of the goddess. This is an allusion to a legend that surrounds Sekhmet, when this goddess almost destroyed humanity. By the Fifth Dynasty, roughly beginning in the 25th century EC, the god known as Ra had become a major god in the Egyptian religion, often associated with the sun. Sekhmet was the daughter of Ra and had the appellation of "eye of Ra," which meant that she was an instrument of the sun god's vengeance. According to George Hart in the work entitled *A Dictionary of Egyptian Gods and Goddesses,* there is a certain myth about mankind threatening to end the reign of Ra's rule on the earth. Ra sends Sekhmet to destroy these foolish mortals who conspired against him. As Ra's instrument of vengeance, Sekhmet metes out divine punishment and destroys the offenders in a nearly orgasmic feeding frenzy. In the myth, Sekhmet's blood-lust was not quelled at the end of battle and this led to her destroying almost all of humanity by drinking human blood like an insatiable vampire. Sekhmet prowled the land, this insatiable goddess slaughtering every human she came across, killing men, women, and children without discretion. Her blood thirst had no limits, and she almost eradicated the whole of humankind. To stop this uncontrollable onslaught of blood-lust, Ra poured out beer dyed with red ochre or, in some versions, by using the oxide mineral known as hematite, into the Nile River so that the waters resembled blood. Mistaking the beer for the blood for which she thirsted, Sekhmet began to lap up the waters and became so drunk that she gave up the slaughter and returned peacefully to Ra.

This festival is probably related to a ritual ceremony held to avert the river flooding during the inevitable inundation at the beginning of every year, when the Nile River ran with red silt that remarkably resembles the color of blood. This red flooding also has a direct connotation to the Old Testament plague where God turns the Nile to blood because

Pharaoh disobeyed the Divine. Thus the monotheistic God of the Hebrews is showing his power and dominance over the pantheon of the Egyptian gods. However, if I may speculate more on the myths that surround the goddess Sekhmet, could it be at least possible that Sekhmet is the cultural memory of humans being preyed upon by their own kind? I elucidated on the idea of the archetype of the werewolf forming from the folk memory of marauding clans wearing the skins of wolves in my book entitled *On Dogman: Tracking the Werewolf Through History*. Could a desert-dwelling tribe, whose totem was the lion, hunted the early tribes of the Nile before they formed a cohesive society? We know for a fact that many cultures wore the skins of the animal whose strengths and cunning they wanted to emulate. A certain warrior class of the Vikings wore bear skins into battle. In the Old Norse, the word for bear is *ber* and the word for coat is *serkr*. Thus these fighters were called the berserker, from which term we get our word berserk. Was there a culture in Egypt that had a similar ideology about the transformative power of wearing the skins of an animal, in this specific case, the hide of the lion? Lions were, after all, fierce and skilled predators. In a land such as Egypt, the idea of at least seasonal cannibalism carried out by an ostracized people, relegated for whatever reason to the desert periphery of the tribal collectives, makes anthropological sense to me. Especially in the light of the ritual held after a war was fought hoping to appease Sekhmet so that the killing and barbarism would stop. This may be a folk memory of the celebration after the defeat of a marauding cannibalistic tribe shortly before the onset of the Old Kingdom period. Our Egyptian vampire goddess may indeed have her roots in the folk memory of actual events.

As all great civilizations that have rose up in the annals of history, the glorious culture of Egypt so too would eventually fall. With the collapse of the New Kingdom in 1070 BC many foreign kings took over and occupied Egypt. The Babylonians, Libyans, Nubians, and later the armies of Alexander the Great marched across Egypt, bringing with them their own legends and beliefs and blending Egyptian gods with their own. The Ptolemys, named after the Greek general Ptolemy, a general of Alexander the Great, ruled Egypt until Rome conquered his heir, Cleopatra. The Romans thus conquering Egypt had thousands of years of beliefs mingling with their culture. Therefore, Greek, Babylonian and other customs were adopted and combined with the older philosophical beliefs of Egypt. In this synchronization, the vampire began to morph once again and take on a more recognizable shape in this Greco-Roman world.

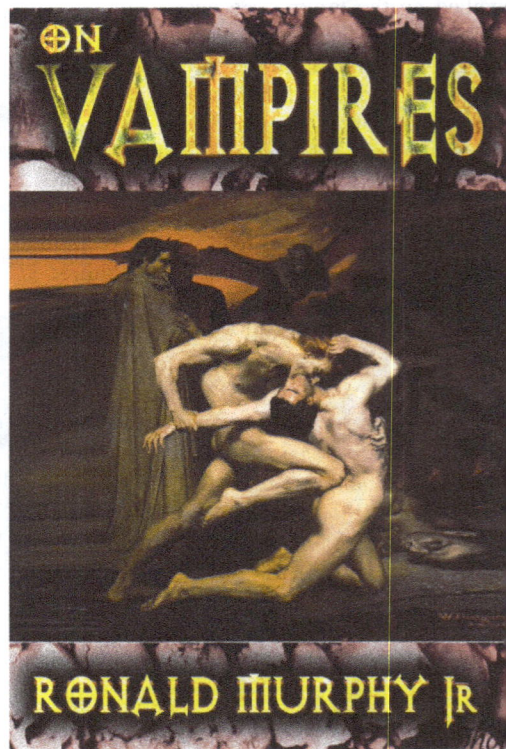

ON
VAMPIRES

RONALD MURPHY JR

KEEPING TRACK(S)

Lars Thomas

Earlier this year, we received this photograph from Ellie Glover, a lady in County Leitrim, Ireland. It was found by her daughter who wondered whether it was from a big cat. There have, apparently, been reports of mystery cats in the area. So, I sent it to my old friend and colleague Lars Thomas and asked him for an identification. He said that it was from a dog and because

this is something which happens to us quite regularly, I decided that it would be a good idea to ask Lars to provide us with a definitive guide to how to identify big cat paw prints. JD

Being able to distinguish between the tracks of cats and the tracks of dogs is one of the most basic pieces of equipment in any self-respecting cryptozoologist's bag of tricks. I am actually being a little frivolous here because, although basic, it is in fact far from easy. For starters, there is an enormous size variation in the actual tracks, from tiny chihuahuas and dainty little housecats with paws the size of 20p coins, to huge mastiffs and big cats, where paw-size in extreme cases is measured in dinnerplates. If you the add variations due to age, soil, weather, humidity, age and activity of the animal in question into the equation, the scope for confusion is huge. However, there are a few important features to look for, which, at least in some cases, makes it possible to identify the animal with some degree of certainty. But it is important to realize from the start that in quite a lot of cases, it will simply not be possible to ID the animal with any reasonable degree of certainty.

Claws

The most obvious difference between cats and dogs, in the footwear department, is in the claws. They both have them of course, but the cats have retractable claws, and the dogs do not. Which means, that under normal circumstances, a dog will leave traces of its claws in the tracks, whereas a cat will not. A dog claw is rather large, straight and blunt, and

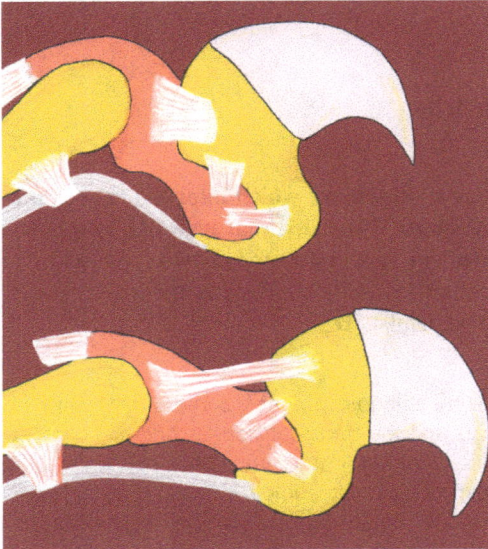

from the tip of the claws and nothing more.

The pads

Both cats and dogs have five pads under each paw – four toe-pads and one large heel-pad. As a very broad and general rule, dog tracks have a tendency to be longer than wide, whereas it is the other way around with cats, but this is nothing more than a general guideline – there are a lot of exceptions. But help may be close at paw, as it were.

If the claw-marks are inconclusive, the large central pad of the tracks may be a help. In cats the back edge is distinctly three-lobed, whereas in dogs, the back-edge curves inward or is straighter than in cats, which gives the pad a more pronounced triangular shape.

And then, of course, there are all the cases where the tracks are too indistinct to be of any help. Then you have to analyse the circumstances of the tracks; how, where and when they were found – and hope for the best. Or search the tracks for embedded hairs or claw sheaths, all of which can be analysed and give an ID of the animals. But that is a whole other kettle of fish.

will leave a distinct triangular mark somewhat similar in shape to a gothic church window. Cats do occasionally leave claw marks – for instance if the ground has been especially soft, and the animal has been jumping. But as the claws of a cat are scimitar-shaped, you don't get the gothic church window as in dogs, but a rather straight sided and very narrow furrow, with a deeper hole at the tip made by the tip of the claw. Sometimes you only get "pinpricks"

COUGAR

—3½ in.—

Front

3 in

Hind

Stride 20–28 in.

DOG

THE TRAGIC STORY OF THE BLUE MONKEY PUB, PLYMOUTH

In May 2017, I read in a book called 'British Inn Signs and Their Stories` by Eric R. Delderfield (David and Charles, Newton Abbot, 1972) a story about a pub with the extraordinary name of 'The Blue Monkey` in Plymouth. There is a double tragedy surrounding the *Blue Monkey* pub, or the former *Blue Monkey* as it burnt to the ground in 2000. The first tragedy is its sad demise into a pile of smoking ashes. The second tragedy is that cryptozoology has lost, perhaps

Fig 1 The Blue Monkey c. 1939
From Plymouth Through Time Derek Tait (2013)

Richard Muirhead and Bob Skinner

forever the interesting possibility that this blue monkey may have been something unique and interesting. However, these misfortunes are more than made up for by the circumstances surrounding the whole story which include the whole folklore surrounding Britain's pubs and their names; such *as The Golden Butterfly*, *The Crocodile*, *The Lamprey* (Gloucester), *The Griffin* (Hertfordshire),*The Golden Tiger*, *The Cat and Custard Pot* in Paddlesworth, Kent and even *The Drunken Duck* in Ambleside, Cumbria!

Plymouth is no stranger to mystery beasts. In July 2015 a photograph was taken of a large cat-like animal there. At the beginning of March 2017 a rare "crocodile-shark" was washed ashore on a beach near the city. As far as monkeys are concerned, Britain has its fair share of escaped simians.

Fortean Times number 43 (Spring 1985) included a round-up of stories of monkeys on the loose:

- In early June 1984 two rare Indonesian silver leaf monkeys escaped from an animal sanctuary in Combe Martin, Devon.
- On August 1st 1984 a male rhesus monkey was first spotted on the outskirts of Frome, Somerset, "bounding across a road."
- About a week before the above-mentioned incident with the rhesus monkey a similar sized Capuchin monkey escaped from

the Vale Garden Centre, in Castleton, between Cardiff and Newport, South Wales.
- Finally, on October 19th 1984, three chimps fled from their cage at *Colchester Zoo* Essex. [1]
- In mid-December 2015, a monkey escaped from *Howletts Wild Animal Park* near Canterbury.

"*The Blue Monkey* was built in 1798 and originally called *The Church Inn*. It was on Crownhill Road it was torched in 2000 and demolished in 2007. The Inn was owned/run by the Octagon Brewery." [2] The Little Book of Devon by John Van der Kiste (2011) says:

"Blue Monkey, Plymouth. Previously known variously as Church Inn, St Bude Inn, St Budeaux Inn, and even Ye Old Budeaux Inn, its name was changed in about 1939. Some say that it was because somebody had seen a monkey climbing on the roof, others that it was in honour of the boys who packed the guns with powder at the Battle of Trafalgar and were left with a pale blue residue on their hands. Towards the end of its days it developed a bad reputation, closed and was put on the market, but failed to sell. It was partly destroyed in an arson attack, and in 2007 the remains were demolished." [3] The Battle of Trafalgar was in October 1805. One web site said the first landlord of The Blue Monkey saw a monkey escaping from a shed. Whilst writing this essay I had a conversation with a friend in my favourite pub in Macclesfield, the

Snowgoose who told me he had visited the Blue Monkey in about 1997 and the landlord at the time had connections with the navy.

The interior of the pub had a cosy "olde-worlde" type atmosphere redolent of some British pubs with trinkets, ornaments and old pictures on the wall. The Dictionary of Pub Names says: "Blue Monkey St Budeaux , Plymouth. Named after the mascot of the Marine Regiment which was stationed nearby in the nineteenth century. It was dressed in a handsome blue monkey-jacket (the short close-fitting jacket worn by seamen.") [4] There was also a Blue Monkey village pub in Kingsand , on the Rame Peninsula on the west side of the River Tamar estuary, whilst Plymouth is on the east. "Blue monkey is a beautiful Grade II 18th century house that has been lovingly refurbished…Blue Monkey was formerly the village pub known as "The Blue Anchor" and was regularly visited by Lord Nelson who could often be seen dancing the night away up in the first floor room. Its name hails partially from this period but also from the time when the house was used as the cannon ball store supplying the fort; "Monkey" was the name given to the triangular formation the cannon balls made when they were stacked, hence the name Blue Monkey" [5] There are some lyrics of a song by 1980s band *Big Country* which go " a mile high the turbines turned, the stokers sweat, the monkeys burned…" (Close Action by Big Country)

The possibility that because there is entirely a naval connection to the Blue Monkey name and no cryptozoology link whatsoever, can be discarded as there was, around 1850, a pub called *The Blue Monkey* in Rotherfield Peppard Parish in Oxfordshire, well inland. This pub became known as the Manor House on the edge of Peppard Common. There is or was also a *Blue Monkey* pub in Nottingham.

There are a number of monkey species called blue monkies. The Eponym Dictionary of Mammals by Bo Beolens and Michael Grayson contains the following information:

- *Cercopithecus mitis boutourlinii*
      Boutourlinii`s blue monkey
- *Cercopithecus mitis mitis*
      Pluto monkey
- *Cercopithecus mitis schoutedenii*
      Schouteden`s blue monkey
- *Cercopithecus mitis stuhlmanni*
      Stuhlmann`s monkey

The image opposite from Wikipedia is of the Stuhlmann`s monkey in the Kakamega Forest, Kenya.

There is a possibility that a menagerie passed through Plymouth at the end of the 1700s and a blue monkey or monkeys escaped and were found by the first landlord of The Blue Monkey pub. In April 1792 Lieutenant Henry Ball anchored his ship in Plymouth Sound with three kangaroos on board, several others had been eaten during the voyage. [6]

Fig 2 Blue Monkey Circopithecus mitis stuhlmanni Sharp photography Charles-jsharp Wikipedia Commons

On November 15th 1838 a rhinoceros was shown in Plymouth according to the Plymouth and Devonport Weekly Journal. A Little White Heron was shot near Kingsbridge, Devon in October 1805, " Temminck...does not doubt that it had escaped from some menagerie, as it is not understood to be even a European bird." [7] Hancock`s Fair and Menagerie had a winter`s store in the Plymouth Refinery site from the 1890s onwards [8] The famous (or should that be infamous if you disapprove of animals being used to entertain humans)

84

"Charles Jamrach moved to London and took over that branch of the business after his father's death in circa 1840. He became a leading importer with agents in other major British ports, including Liverpool, Southampton and Plymouth." [9]

Here is a photograph kindly supplied to me (Richard) by Macclesfield local historian and bookseller Ann-Marie Pond. It shows Sir Philip Brocklehurst (from whose private zoo the wallabies which later populated The Roaches in the Peak District escaped just before World War 2) holding what looks like a baboon. The photo is from either 'Around Rushton' or 'Around the Dane Valley Gradbach to Bosley' both written by Sheila Hine and published by Churnet Vally Books

"The blue monkey or diadem monkey is a species of guenon native to parts of east, central and southern Africa including the Congo River basin. The Blue Monkey is found in rain forests and montane bamboo forests, and lives largely in the forest canopy, coming to the ground infrequently.

It is very dependent on humid, shady areas with plenty of water. Despite its name, the Blue Monkey is not noticeably blue. It has little hair on its face and this does sometimes give a blue appearance, however, it never has the vivid blue appearance of a mandrill, for example.

The Blue Monkey's fur is short and mainly a grizzled brown colour apart

Fig 3 Sir Philip Brocklehurst with pet baboon.

from the face (which is dark with a pale or yellowish patch on the forehead – the 'diadem' from which the species derives its common name) and the mantle, which varies between subspecies. Typical sizes are from 50 to 65 centimetres in length (not including the tail, which is almost as long as the rest of the animal), with females weighing a little over 4 kilograms and males up to 8 kilograms.

Blue Monkeys are catarrhine – the nostrils are close together and they face downward. The nail on each digit is flattened and the thumb is opposable." [10] Strange to say, an advertisement in the United States Telegraph (Washington D.C.) of August 23rd 1828 for the Boston Caravan of Living Animals mentioned a "blue monkey from India." I could find no information on the Net for a blue monkey in India. There is of course the blue painted Hindu god Hanuman. There was also a (blue) monkey jacket. According to Wikipedia: "A monkey jacket is a waist length jacket tapering at the back to a point.

Historically monkey jackets were commonly worn by sailors." [11]

The web site softschools.com states that blue monkeys live in large groups (called troops) composed of 10 to 40 members.

Fig 4. Photo of a swabbie in a monkey jacket, circa mid to late 19th century. American Civil War. Unknown date. The Navy & Marine Living History Association. Wikipedia Commons.

REFERENCES

1.   Fortean Times no 43 Spring 1985
2.   E-mail from Trisha White , Plymouth and West Devon Record Office May 10th 2017th 2013 George Wombwell and Travelling Menageries blog http:// www.georgewombwell.com/ gw_blog/?p=223
3.   Animal Corner https://
     animalcorner.co.uk/animals/ gold-silver-blue-monkeys/
4.   Wikipedia. Monkey jacket. https://en.wikipedia.org/wiki/ Monkey_jacket
5.   John Van der Kiste The Little Book of Devon (2011)

Fig 4 The Blue Monkey after its destruction by fire
© Mike Lobb http://www.geograph.org.uk/reuse.php?id=915044

6.   Dictionary of Pub Names Wordsworth Books (2006) p. 51

7.   Blue Monkey   http://luxurycornwall.com/10-dog-friendly-holidays-in-cornwall-and-devon/

8.   Menagerie: The history of exotic animals in England Caroline Grigson O.U.P 2016

9.   Transactions of the Plymouth Institution and Devon and Cornwall Natural History Society 1830 pp 323-324.

10. See Plymouth http://www.mawer.clara.net/loc-plym.html

11. Jamrach`s Animal Emporium August 8

Reviews

- **ISBN-10:** 193839870X
- **ISBN-13:** 978-1938398704

In his previous book, Lyle Blackburn examined the infamous beast of Boggy Creek, the ape-like monster that has haunted the swamps around Fouke, Arkansas for decades. In this, his second volume on mystery primates Lyle widens his scope, examining reports from across the South-East USA.

The Beast of Boggy Creek showed that Lyle was both a first-rate researcher and a masterful writer. I eulogized over the book in these very pages and still hold it to be one of the best volumes written on the sasquatch. His second book does not disappoint and is every bit as gripping and informative as his first.

What emerges in *Beyond Boggy Creek* is a picture of a population of swamp dwelling creatures of whose nature seems far more aggressive than the more familiar bigfoot of the northern states and Canada. Some of the reports and encounters in the pages read like horror stories. Take the case of Mike Woosley, hunting from a deer stand in Sabine Parish Game Reserve, western Louisiana in December of 1981. Having seen an exhausted doe emerge from the forest he assumed it was being chased by a buck and got ready to shoot the male deer as it appeared. What came out of the forest however was a hair-covered, muscular, upright creature over seven feet tall.

Thinking it was some kind of prank, Woosley looked more closely at the figure through the scope of his gun and realized to his horror that the creature was real. The movement caught the beast's eye. It began to vocalize in an aggressive manner and was

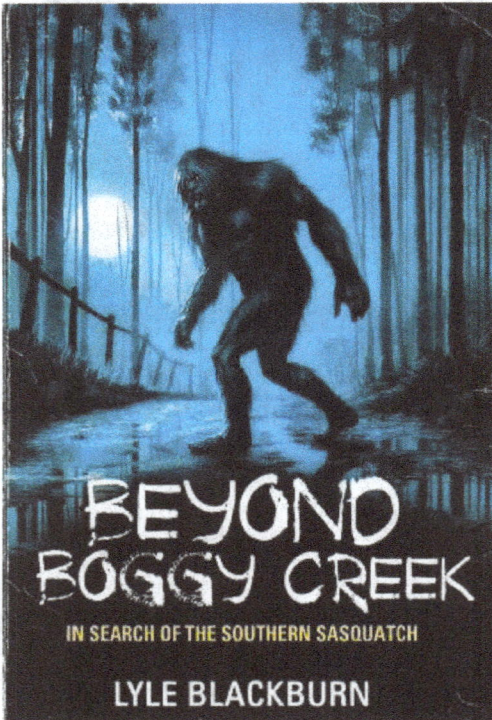

- **Paperback:** 310 pages
- **Publisher:** Anomalist Books (1 Feb. 2017)
- **Language:** English

answered by a second in the forest. The two creatures pursued the terrified hunter for half a mile back to his truck. One was chasing him, the other flanking him in the forest. He only just made it back to the vehicle by turning and firing at one of the pursuing creatures, buying himself enough time to get into the truck and drive away.

In another case from 2000, Tim Humphreys and his family were driven out of the farm in the KIamichi Mountains of Oklahoma by 8 foot, foul smelling, ape-like beasts. The creatures had stolen meat from a storage freezer and tried to enter the house. They beat on the sides of the house, peered through windows and twisted the knob off the back door.

Even more creepy is the tale of a couple driving near Lake Crook, near Paris, Texas one evening in 2002. After getting stuck in the mud (without a mobile phone) the man walked to the nearest town to get help. The woman stayed in the car. She dozed off but was awakened by a strange scream. She saw a trio of hairy creatures that seemed both simian and human-like. One was male and the other two, females. The male reached through the open window and grabbed the woman's leg. It tried to haul it's screaming victim out of the car as she hung onto the steering wheel. Finally, she grabbed a soda bottle and hit the creature's arm. It let go long enough for her to wind up the window and lock the door. The creatures retreated and soon her husband returned with help. Her leg was bleeding from scratches and she was taken to a hospital in Paris.

Other stories include a hiker pursued for over 24 hours by two ape creatures hurling rocks, and cases of the swamp apes

attempting to and possibly succeeding in snatching human children. In recent years, it has become apparent that many people are going missing in national parks and forests in the USA and Canada under very weird circumstances. The disappearances are not consistent with bear, wolf or puma attacks. This, taken with the reports of aggressive sasquatch that may now be looking at humans as a source of food raises disturbing possibilities.   RF

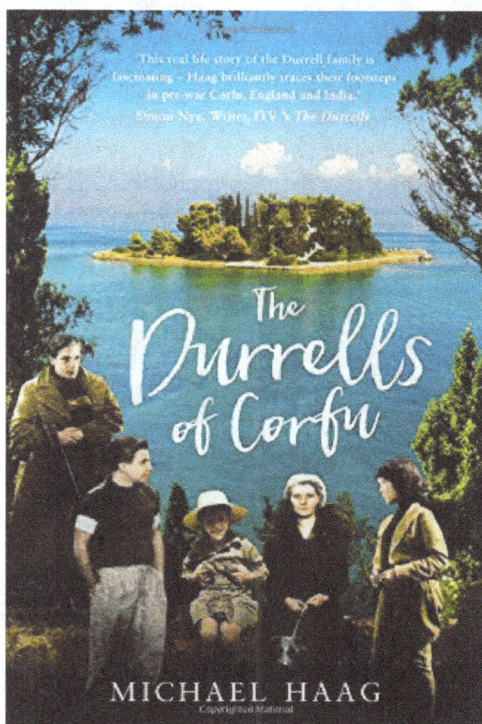

- **Paperback:** 224 pages
- **Publisher:** Profile Books; Main edition (20 April 2017)
- **Language:** English
- **ISBN-10:** 1781257884

- **ISBN-13:** 978-1781257883

We live in a world of conspiracy theories, where no statement of supposed fact goes unchallenged, and for every book or essay asserting the truth if a historical supposition, there are several more suggesting the exact opposite. As a researcher into Fortean subjects (to use the term at its very broadest) I have often found myself in this very position of becoming one of those people who seem to spend an awful lot of the time asking inconvenient questions.

Now, it us a matter of record that one of my great heroes is author and conservationist, the late Gerald Durrell. And - I suspect like a lot of people - I was introduced to his work, by his phenomenally successful 1956 book *My Family and other Animals*. The book is an autobiographical account of five years in his childhood on the Greek island of Corfu. Gerald Durrell, is age 10 at the start of the saga, which tells of his family, pets and life during a sojourn on the island of Corfu.

Over to those jolly nice people at Wikipedia:

"The book is divided into three sections, marking the three villas where the family lived on the island. Gerald is the youngest in a family consisting of their widowed mother, the eldest son Larry, the gun-mad Leslie, and diet-obsessed sister Margo together with Roger the dog.

They are fiercely protected by their taxi-driver friend Spiro (Spiros "Americano" Halikiopoulos) and mentored by the polymath Dr. Theodore Stephanides who provides Gerald with his education in natural history. Other human characters, chiefly eccentric, include Gerald's private tutors, the artistic and literary visitors Larry invites to stay, and the local people who befriend the family."

It is one of the books upon which I have based the way I have spent my life, and I still remember how disappointed I was when I found out that it wasn't actually Gospel truth. (The fact that neither are the Gospels - probably - didn't actually upset me so much.)

This is the first book that I have read which actually tries to sift out the truth from the fiction in the actual history of the Durrell family's sojourn on Corfu. Other books have partially covered the subject before, most notably Douglas Botting's biography of Durrell, but this is the first book to solely devote itself to the subject.

Various people have criticised Durrell for having left several of the major characters from his childhood out of the books. Larry Durrell's first wife Nancy, Theodore's wife Mary and their daughter Alexia are completely excised from the narrative, despite the fact that Alexia was the young Gerry's closest friend on the island, and is - coincidentally - the only person from those days still living (she celebrated her 90th birthday this year).

Durrell has been retrospectively accused of various sins of misogyny by leaving out such pivotal characters, who all happen to be female.

But it seems that these charges are unfounded, and that the real reasons that they were excised from his version if history, is that he didn't want his idyllic childhood tale to be sullies by the spectre of two of the major characters getting divorced. These days, an author probably wouldn't be bothered by this, but Durrell was from the upper echelons of British society, and 1956 was a heck of a long time ago.

My Family and Other Animals spawned two full length sequels, and a number of short stories, and has been adapted for television three times, most recently in a series called *The Durrells* which has just finished its second series on ITV in the UK.

This series has come in for some criticism because it strays quite considerably from the narrative in the three books, although I had already guessed (from reading Douglas Botting) that in many ways it actually told a story nearer to the truth than the three books or the previous TV adaptations.

This new TV series proved to be very redemptive to Margaret (Margo) Durrell, and particularly to her brother Leslie. The former is portrayed in the books as being an airhead, with a string of comical boyfriends and a tendency towards obesity and acne.

When the books were first published, Botting describes how she complained that Gerry had portrayed her in a cartoonish character, and ignored "the strength of character" she believed she had. Both this book and The Durrells redeems her, showing an intelligent and insightful (though still, slightly shallow) young woman.

But, with the benefit of 82 years of hindsight, it was Leslie who suffered the most from his brother's literary prowess. The real Leslie, as shown in The Durrells and in this book, was a talented painter and photographer, and somewhat of a Lothario, whereas in the book he is portrayed as a loutish meathead whose main interest in life was killing things and collecting weapons.

He was not even forty when *My Family and Other Animals* was first published, but from then on his life seemed to go downhill fairly rapidly. He moved to Kenya with his wife Doris, but their business failed and they returned to Britain in 1968 whereupon he was accused/charged/convicted (depending on which account you believe) of fraud, and when he died of heart failure in the 1980s, he was estranged from all of his family.

It would seem that he was somewhat unstable already. Konstantina Konshevin writes:

"Leslie returned to England with his mother, Gerald, Margaret and the family's Corfiot maid, Maria Kondos when the Second World War broke. (Margaret, of course, soon afterwards went back 'home' to Corfu.) The Durrells settled in Bournemouth and Leslie tried to enlist in the army but was rejected on the grounds of ill-health, something that was a setback for him. Instead, he spent the war working in an RAF factory.

Shortly after the family returned to England, Leslie had a brief romance with the family's Corfiot friend and live-in maid, Maria Kondos, that produced a son, Anthony. However, the romance was short-lived."

But reading this remarkable book has given me a horrible feeling.

Could Gerald's book, which has given so much joy to so many people over the years, have been a significant causative factor in the decline of his brother, who in his own way was as talented as the rest if his family, but who was never fated to enjoy the fruits of his talents.

It is a sad and disquieting thought.

MUIRHEAD'S MYSTERIES

Volume One

Richard Muirhead

"Harvested from his regular, long-running online column 'Muirhead's Mysteries' contained within the *CFZ Bloggo*, in this fascinating volume – which I sincerely hope will be just the first of many in a continuing 'MM' series – Richard has surveyed and selected a truly eclectic array of cryptozoological and other animal-related enigmas with which to bemuse and bedazzle his readers."

Dr Karl Shuker

Typeset by a bunch of chickens

STILL ON THE TRACK OF UNKNOWN ANIMALS

The Centre for Fortean Zoology, or CFZ, is a non profit-making organisation founded in 1992 with the aim of being a clearing house for information, and coordinating research into mystery animals around the world.

We also study out of place animals, rare and aberrant animal behaviour, and Zooform Phenomena; little-understood "things" that appear to be animals, but which are in fact nothing of the sort, and not even alive (at least in the way we understand the term).

Not only are we the biggest organisation of our type in the world, but - or so we like to think - we are the best. We are certainly the only truly global cryptozoological research organisation, and we carry out our investigations using a strictly scientific set of guidelines. We are expanding all the time and looking to recruit new members to help us in our research into mysterious animals and strange creatures across the globe.

Why should you join us? Because, if you are genuinely interested in trying to solve the last great mysteries of Mother Nature, there is nobody better than us with whom to do it.

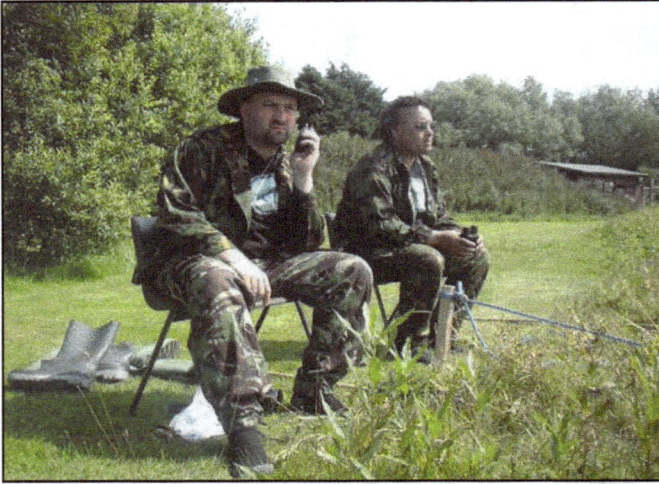

Members get a four-issue subscription to our journal *Animals & Men.* Each issue contains nearly 100 pages packed with news, articles, letters, research papers, field reports, and even a gossip column! The magazine is Royal Octavo in format with a full colour cover. You also have access to one of the world's largest collections of resource material dealing with cryptozoology and allied disciplines, and people from the CFZ membership regularly take part in fieldwork and expeditions around the world.

The CFZ is managed by a three-man board of trustees, with a non-profit making trust registered with HM Government Stamp Office. The board of trustees is supported by a Permanent Directorate of full and part-time staff, and advised by a Consultancy Board of specialists - many of whom are world-renowned experts in their particular field. We have regional representatives across the UK, the USA, and many other parts of the world, and are affiliated with other organisations whose aims and protocols mirror our own.

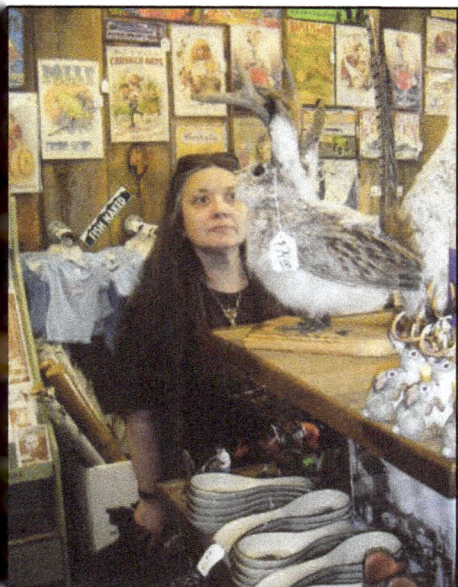

You'll find that the people at the CFZ are friendly and approachable. We have a thriving forum on the website which is the hub of an ever-growing electronic community. You will soon find your feet. Many members of the CFZ Permanent Directorate started off as ordinary members, and now work full-time chasing monsters around the world.

Write to us, e-mail us, or telephone us. The list of future projects on the website is not exhaustive. If you have a good idea for an investigation, please tell us. We may well be able to help.

We are always looking for volunteers to join us. If you see a project that interests you, do not hesitate to get in touch with us. Under certain circumstances we can help provide funding for your trip. If you look on the future projects section of the website, you can see some of the projects that we have pencilled in for the next few years.

In 2003 and 2004 we sent three-man expeditions to Sumatra looking for Orang-Pendek - a semi-legendary bipedal ape. The same three went to Mongolia in 2005. All three members started off merely subscribers to the CFZ magazine. Next time it could be you!

We have no magic sources of income. All our funds come from donations, membership fees, and sales of our publications and merchandise. We are always looking for corporate sponsorship, and other sources of revenue. If you have any ideas for fund-raising please let us know. However, unlike other cryptozoological organisations in the past, we do not live in an intellectual ivory tower. We are not afraid to get our hands dirty, and furthermore we are not one of

those organisations where the membership have to raise money so that a privileged few can go on expensive foreign trips. Our research teams, both in the UK and abroad, consist of a mixture of experienced and inexperienced personnel. We are truly a community, and work on the premise that the benefits of CFZ membership are open to all.

Reports of our investigations are published on our website as soon as they are available. Preliminary reports are posted within days of the project finishing.

Each year we publish a 200 page yearbook containing research papers and expedition reports too long to be printed in the journal. We freely circulate our information to anybody who asks for it.

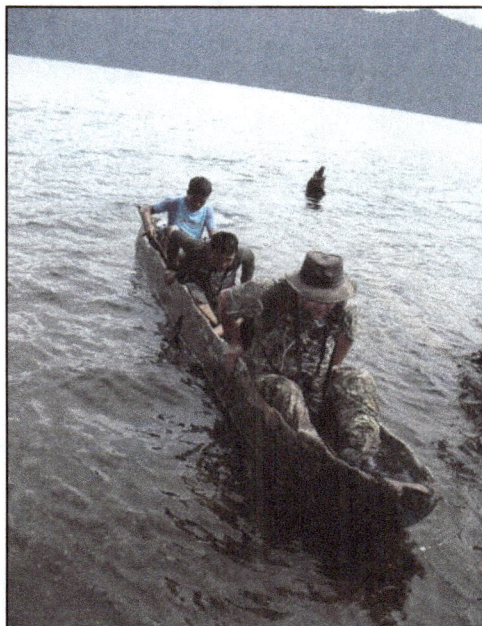

We have a thriving YouTube channel, CFZtv, which has well over two hundred self-made documentaries, lecture appearances, and episodes of our monthly webTV show. We have a daily online magazine, which has over a million hits each year.

Each year since 2000 we have held our annual convention - the Weird Weekend. It is three days of lectures, workshops, and excursions. But most importantly it is a chance for members of the CFZ to meet each other, and to talk with the members of the permanent directorate in a relaxed and informal setting and preferably with a pint of beer in one hand. Since 2006 - the Weird Weekend has been bigger and better and held on the third weekend in August in the idyllic rural location of Woolsery in North Devon.

Since relocating to North Devon in 2005 we have become ever more closely involved with other community organisations, and we hope that this trend will continue. We have also worked closely with Police Forces across the UK as consultants for animal mutilation cases, and we intend to forge closer links with the coastguard and other community services. We want to work closely with those who regularly travel into the Bristol Channel, so that if the recent trend of exotic animal visitors to our coastal waters continues, we can be out there as soon as possible.

Apart from having been the only Fortean Zoological organisation in the world to have consistently published material on all aspects of the subject for over a decade, we have achieved the following concrete results:

- Disproved the myth relating to the headless so-called sea-serpent carcass of Durgan beach in Cornwall 1975
- Disproved the story of the 1988 puma skull of Lustleigh Cleave
- Carried out the only in-depth research ever into the mythos of the Cornish

Owlman.

- Made the first records of a tropical species of lamprey
- Made the first records of a luminous cave gnat larva in Thailand
- Discovered a possible new species of British mammal - the beech marten
- In 1994-6 carried out the first archival fortean zoological survey of Hong Kong
- In the year 2000, CFZ theories were confirmed when a new species of lizard was added to the British List
- Identified the monster of Martin Mere in Lancashire as a giant wels catfish
- Expanded the known range of Armitage's skink in the Gambia by 80%
- Obtained photographic evidence of the remains of Europe's largest known pike
- Carried out the first ever in-depth study of the ninki-nanka
- Carried out the first attempt to breed Puerto Rican cave snails in captivity
- Were the first European explorers to visit the `lost valley` in Sumatra
- Published the first ever evidence for a new tribe of pygmies in Guyana
- Published the first evidence for a new species of caiman in Guyana
- Filmed unknown creatures

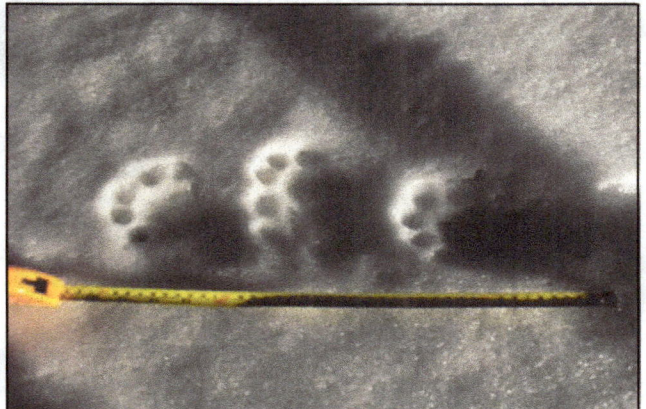

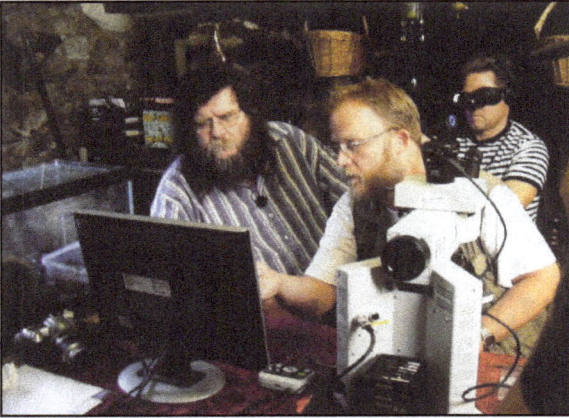

on a monster-haunted lake in Ireland for the first time

• Had a sighting of orang pendek in Sumatra in 2009

• Found leopard hair, subsequently identified by DNA analysis, from rural North Devon in 2010

• Brought back hairs which appear to be from an unknown primate in Sumatra

• Published some of the best evidence ever for the almasty in southern Russia

CFZ Expeditions and Investigations include:

• 1998 Puerto Rico, Florida, Mexico (Chupacabras)
• 1999 Nevada (Bigfoot)
• 2000 Thailand (Naga)
• 2002 Martin Mere (Giant catfish)
• 2002 Cleveland (Wallaby mutilation)
• 2003 Bolam Lake (BHM Reports)
• 2003 Sumatra (Orang Pendek)

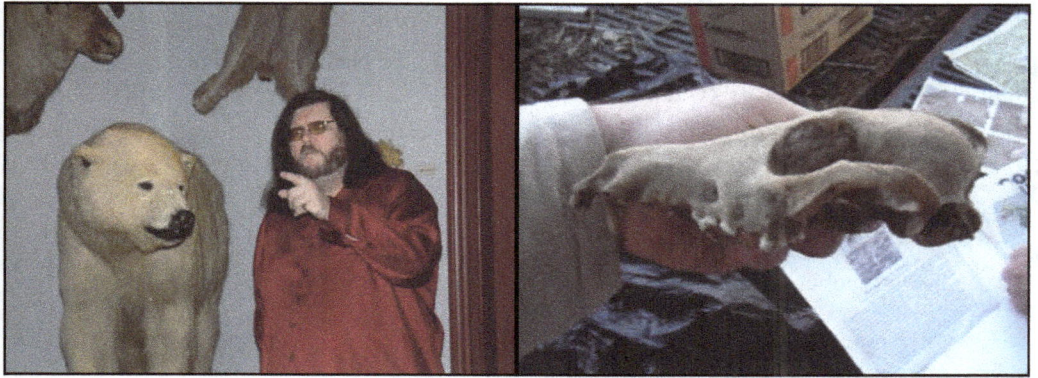

- 2003 Texas (Bigfoot; giant snapping turtles)
- 2004 Sumatra (Orang Pendek; cigau, a sabre-toothed cat)
- 2004 Illinois (Black panthers; cicada swarm)
- 2004 Texas (Mystery blue dog)
- Loch Morar (Monster)
- 2004 Puerto Rico (Chupacabras; carnivorous cave snails)
- 2005 Belize (Affiliate expedition for hairy dwarfs)
- 2005 Loch Ness (Monster)
- 2005 Mongolia (Allghoi Khorkhoi aka Mongolian death worm)

- 2006 Gambia (Gambo - Gambian sea monster , Ninki Nanka and Armitage's skink
- 2006 Llangorse Lake (Giant pike, giant eels)
- 2006 Windermere (Giant eels)
- 2007 Coniston Water (Giant eels)
- 2007 Guyana (Giant anaconda, didi, water tiger)
- 2008 Russia (Almasty)
- 2009 Sumatra (Orang pendek)
- 2009 Republic of Ireland (Lake Monster)
- 2010 Texas (Blue Dogs)
- 2010 India (Mande Burung)
- 2011 Sumatra (Orang-pendek)
- 2012 Sumatra (Orang Pendek)
- 2014 Tasmania (Thylacine)
- 2015 Tasmania (Thylacine)
- 2016 Tasmania (Thylacine)
- 2017 Tasmania (Thylacine)

For details of current membership fees, current expeditions and investigations, and voluntary posts within the CFZ that need your help, please do not hesitate to contact us.

The Centre for Fortean Zoology,
Myrtle Cottage,
Woolfardisworthy,
Bideford, North Devon
EX39 5QR

Telephone 01237 431413
Fax+44 (0)7006-074-925
eMail info@cfz.org.uk

Websites:

www.cfz.org.uk
www.weirdweekend.org

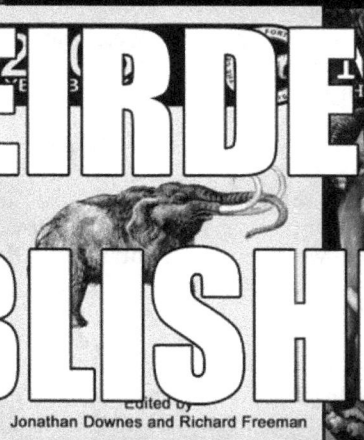

THE WORLD'S WEIRDEST PUBLISHING COMPANY

HOW TO START A PUBLISHING EMPIRE

Unlike most mainstream publishers, we have a non-commercial remit, and our mission statement claims that "we publish books because they deserve to be published, not because we think that we can make money out of them". Our motto is the Latin Tag *Pro bona causa facimus* (we do it for good reason), a slogan taken from a children's book *The Case of the Silver Egg* by the late Desmond Skirrow.

WIKIPEDIA: "The first book published was in 1988. *Take this Brother may it Serve you Well* was a guide to Beatles bootlegs by Jonathan Downes. It sold quite well, but was hampered by very poor production values, being photocopied, and held together by a plastic clip binder.

In 1988 A5 clip binders were hard to get hold of, so the publishers took A4 binders and cut them in half with a hacksaw. It now reaches surprisingly high prices second hand.

The production quality improved slightly over the years, and after 1999 all the books produced were ringbound with laminated colour covers. In 2004, however, they signed an agreement with Lightning Source, and all books are now produced perfect bound, with full colour covers."

Until 2010 all our books, the majority of which are/were on the subject of mystery animals and allied disciplines, were published by `CFZ Press`, the publishing arm of the Centre for Fortean Zoology (CFZ), and we urged our readers and followers to draw a discreet veil over the books that we published that were completely off topic to the CFZ.

However, in 2010 we decided that enough was enough and launched a second imprint, `Fortean Words` which aims to cover a wide range of non animal-related esoteric subjects. Other imprints will be launched as and when we feel like it, however the basic ethos of the company remains the same: Our job is to publish books and magazines that we feel are worth publishing, whether or not they are going to sell. Money is, after all - as my dear old Mama once told me - a rather vulgar subject, and she would be rolling in her grave if she thought that her eldest son was somehow in `trade`.

Luckily, so far our tastes have turned out not to be that rarified after all, and we have sold far more books than anyone ever thought that we would, so there is a moral in there somewhere…

Jon Downes,
Woolsery, North Devon
July 2010

CFZ PRESS

CFZ Press is our flagship imprint, featuring a wide range of intelligently written and lavishly illustrated books on cryptozoology and the quirkier aspects of Natural History.

CFZ Classics is a new venture for us. There are many seminal works that are either unavailable today, or not available with the production values which we would like to see. So, following the old adage that if you want to get something done do it yourself, this is exactly what we have done.

Desiderius Erasmus Roterodamus (b. October 18th 1466, d. July 2nd 1536) said: "When I have a little money, I buy books; and if I have any left, I buy food and clothes," and we are much the same. Only, we are in the lucky position of being able to share our books with the wider world. CFZ Classics is a conduit through which we cannot just re-issue titles which we feel still have much to offer the cryptozoological and Fortean research communities of the 21st Century, but we are adding footnotes, supplementary essays, and other material where we deem it appropriate.

http://www.cfzpublishing.co.uk/

Fortean Words is a new venture for us. The F in CFZ stands for "Fortean", after the pioneering researcher into anomalous phenomena, Charles Fort. Our Fortean Words imprint covers a whole spectrum of arcane subjects from UFOs and the paranormal to folklore and urban legends. Our authors include such Fortean luminaries as Nick Redfern, Andy Roberts, and Paul Screeton. . New authors tackling new subjects will always be encouraged, and we hope that our books will continue to be as ground-breaking and popular as ever.

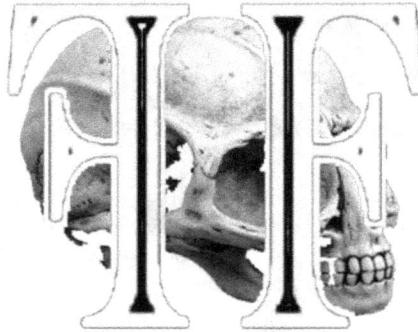

Just before Christmas 2011, we launched our third imprint, this time dedicated to - let's see if you guessed it from the title - fictional books with a Fortean or cryptozoological theme. We have published a few fictional books in the past, but now think that because of our rising reputation as publishers of quality Forteana, that a dedicated fiction imprint was the order of the day.

http://www.cfzpublishing.co.uk/